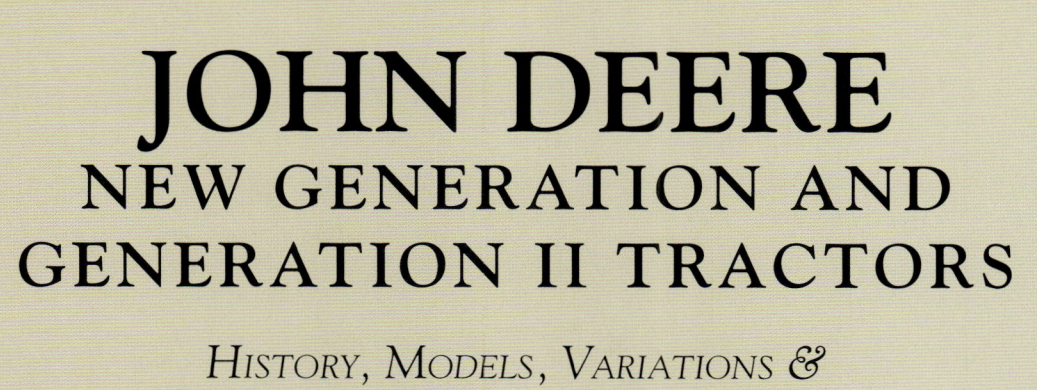

JOHN DEERE
NEW GENERATION AND
GENERATION II TRACTORS

History, Models, Variations &
Specifications 1960s–1970s

2010

JOHN DIETZ

Voyageur Press

First published in 2010 by Voyageur Press, an imprint of
MBI Publishing Company, 400 First Avenue North, Suite 300,
Minneapolis, MN 55401 USA

Voyageur Press titles are also available at discounts in bulk
quantity for industrial or sales-promotional use. For details write
to Special Sales Manager at MBI Publishing Company, 400 First
Avenue North, Suite 300, Minneapolis, MN 55401 USA.

To find out more about our books, visit us online at
www.voyageurpress.com.

Library of Congress Cataloging-in-Publication Data

Dietz, John, 1946-
 John Deere new generation and generation II tractors : history,
models, variations & specifications 1960s-1970s / John Dietz.
  p. cm. — (Tractor legacy series)
  Includes bibliographical references and index.
  ISBN 978-0-7603-3600-7 (plc)
  1. John Deere tractors—United States—History. 2. Farm
tractors—United States—History. I. Title.
  TL233.6.J64D53 2010
  629.225'2—dc22
  2009033377

Editors: Lee Klancher and Margret Aldrich
Design Manager: LeAnn Kuhlmann
Layout by: Kazuko Collins

On the frontis: Most of Deere's 2520 tractors were for sugar cane
growers who needed the high clearance and power for cultivating.
Bruce Keller

On the title pages: The John Deere 2010 and 1010 were the
smallest New Generation tractors introduced in 1960. *Tony Gerber*

Printed in China

CONTENTS

John Deere General-Purpose Tractors brochure, 1950s.

FARM MACHINERY MAKERS IN THE 1950s

In our town in 1953, if you were well-behaved and quiet, the local pharmacist would let you sit on the oiled fir floor and read comics as long as you liked. You had to put them back neatly, of course. At that time, in rural small-town America, every household seemed to have new postwar children. We later came to know ourselves as Baby Boomers.

New schools were built and portable classrooms were about to be invented. The men in uniform had come home to become husbands, workers, merchants, shopkeepers, officials, and farmers. The genius and commitment they'd put into winning wars overseas in Europe and on the Pacific Ocean were redirected. They were building a powerhouse of a system, an economy never before seen in world history. The material core of that new economy was the extraction and use of natural resources for the generation of wealth. While these men were building a new way of life, young boys like me sat on wooden floors reading about legendary characters and comic-book heroes. My favorites were Paul Bunyan and Superman.

Another comic-like book was popular back then. Had I found it, it would have been a keeper. The cover was a larger-than-life painting of a boy on a green tractor. He was wearing denim coveralls, like the kind I'd worn to my first day of kindergarten, and a plaid shirt. He was built strong, with broad shoulders. His sleeves were rolled up to the elbow while he firmly gripped a three-spoke steering wheel at the 10-and-2 position. His light brown hair was tousled by the breeze. He had a broad, healthy grin, and the sky above him was a mixture of blue and fluffy clouds. At his right knee, he could pull or push a long lever coming out of the tractor floor. In front of his knee was a big electric light, a headlight. In a patch of exploding bright blue text at the bottom right corner of the cover were words of power and excitement foretelling the action inside: "New A and B Series." The title was *John Deere General-Purpose Tractors*. The brightest thing on that cover (besides the youth's smile) were the bright yellow words "JOHN DEERE." I wasn't likely to ever be Superman, but given another eight or ten years, I could be just like that youth on the tractor.

Our generation didn't know what had happened before we were born. We didn't know of war or the Great Depression. We didn't know what it was like to go to bed by lamplight, and we didn't know the connection between horses and work. We knew the sound of the radio, and valued how it connected us to faraway places. We heard rumors of a "tele-vision," a device combining radio, newspapers, and movies in a box you could watch in the living room. In the cities we saw evidence

Farm dealerships were modernizing in the 1950s. This is the knotty pine parts counter at the Fred Haar Company dealership in Freeman, South Dakota, as it appeared around 1955. *Al Haar*

of wealth and progress: the last streets were being paved, and most families had a big six-seat car fully equipped with a radio. Factories were at full capacity, and my hometown's shipyard was busy building huge ore carriers for the Great Lakes. With a five-digit phone number, we could talk to anyone else in the city. With a half-hour drive through the rolling country and two villages, we could get to Grandma's house for Sunday dinner and visits with cousins. The road curved around hillsides, rising and falling, crossing bridges and cutting through dotted woodlands. We'd see dairies and farms along the way. We might see a tractor. Sometimes we made a game of counting telephone poles. We looked for Burma Shave signs. Our new generation lived in a time when things were good in America.

While my father and his brothers soldiered on at factories for low or minimum hourly wages, others reached high for seemingly boundless opportunities, and many succeeded. My 36-page John Deere comic was a product

of collaboration by successful artists, writers, and engineers. That brochure symbolized the efforts of private enterprise to create a vision of a bright new world for a new generation of customers. In 1953, America still had five million farms, and each one of them needed a variety of equipment. Nearly every farm had a tractor, and many had two. But while there had been more than 180 tractor manufacturers at the start of the century, that number had been cut to fewer than 10 that were significant suppliers in this mid-century postwar boom. The leaders of these powerful companies made decisions that determined their market share, gained or lost them wealth, and supplied a nation with the stuff of farming and ranching. Like action heroes, their companies jostled on the pages of the American economy for leadership and power.

Deere & Company in Moline, Illinois, was one of the eight surviving postwar farm machinery manufacturers in North America. Together, these companies built 99.8 percent of all tractors sold during this time. In 1953, Deere manufactured a long line of farm equipment in addition to eight models of their venerable two-cylinder tractors. The company had dealers in every state in the Union and, north of the border, in every province of the Federation.

At those dealerships, farmers could buy a wide range of equipment for tillage (plows, cultivators, disk harrows, disk tillers), for planting (corn, vegetables, potatoes, grain, fertilizer), for haying (mowers, rakes, wagons, balers), and for harvesting (combines; windrowers; threshers; corn pickers, shellers, and snappers; harvesters for cotton, potatoes, and beets) and even grain dryers. Dealerships also supplied equipment for shaping the earth, like dozers, blades, scrapers, scoops, and levelers. In addition, Deere products included a rotary cutter for grass and brush, a manure loader, a manure spreader, elevators for hay or grain, hammer mills for grinding feed, snowplows, and stationary engines.

Deere & Company had competitors down the road, however, who were eager for the same farm business. These were the other survivors of a winnowing process that had been underway from the beginning of tractor power in farming. The faltering farm economy in the

A restored John Deere 4010 row-crop tractor.

1920s had reduced competitors to approximately 30 fairly stable tractor manufacturers. Others dropped off during the Great Depression and World War II. In the first half of the 1950s, ordered by market share, the Big Eight in tractors were: International Harvester, 30.6 percent; Ford, 19.4 percent; Deere, 14.5 percent; Massey-Ferguson, 10.8 percent; Allis-Chalmers, 10.3 percent; Oliver, 5.4 percent; Case, 5.1 percent; Minneapolis-Moline, 3.6 percent.

That winnowing process continued, and more companies were lost in the following decades. It finally stopped in 1991, when AGCO Corporation purchased the White-New Idea Farm Equipment Company. At that point, the surviving full-line farm machinery companies with dealerships were AGCO, Case IH, Ford-New Holland, and Deere. Of course, they weren't

alone in the market. There were other significant tractor manufacturers overseas, but the consolidation in North America was over. These four continue today as the leading suppliers of farm tractors in the United States and Canada.

Among the Big Four in 1991, the only survivor who hadn't gone through a bankruptcy or merger was Deere & Company. The company's survival through the turbulent economics of the Baby Boom era was guided primarily by a man who married the boss's daughter after the war and entered the boardroom in 1953. While Deere & Company grew through the manufacture of products that garnered increasing respect and market share in the 1960s and 1970s, it also became widely recognized as one of the best-managed corporations in North America.

This is the story of that era.

The first and last of the John Deere two-cylinder tractors, 1924 to 1960, as seen in the Brandon, Manitoba, collection of Brent and Gregg Campbell.

DEERE ENGAGES THE WORLD

A New Leader

Corporations have rights and responsibilities. Just like individuals, they are born and they can die. Corporations can be convicted of criminal offenses. They are the outward expression of many internal dynamics. They inhale life-sustaining supplies, release energy, and produce waste. They require leadership, policy, accountability, and connectivity.

In the 1960s and 1970s, Deere & Company was among the healthiest of all the large corporations in North America. The company had a long history. It was founded in 1837 and incorporated in 1868. Its first public share offering was in 1911. In the 1960s, the fifth generation of John Deere descendants controlled the board of directors. Deere survived by efficiently building products the market wanted.

The leader of the corporation from 1955 through 1982 was William Alexander Hewitt. He was a leader without a farming background, a leader from California who had risen in wartime to become a battleship commander, a leader who was raising a young family with John Deere genetics.

Before President Charles Deere Wiman died on May 12, 1955, he weighed the choices for a successor and settled on 40-year-old Bill Hewitt as Deere's sixth president.

It proved to be a great choice. Led by Hewitt, the corporation gained worldwide respect. When he passed on the helm to his successor, Hewitt became the United States ambassador to Jamaica. He died on May 16, 1998, back at home in California. His era is now recognized as the best period in Deere's history. He arrived at a time when great leadership was necessary.

Twins Adrienne and Anna, and younger brother Sandy, cuddle up to their papa, William Alexander Hewitt, around 1965. *Anna Wolfe collection*

Official U.S. Navy photo of Lt. William A. Hewitt, stationed on board the USS California and based in San Francisco, 1944. *Anna Wolfe collection*

Born in San Francisco on August 9, 1914, Hewitt had to work for two years after high school as a bank messenger before he could attend college. He graduated from the University of California, Berkeley, in 1937 with a degree in economics. While there, he blossomed socially and fell in with a group that was destined for success.

His close friends included Robert S. McNamara, Walter Haas Jr., Willard Goodwin, and Stanley Johnson. McNamara became the U.S. Secretary of Defense, president of Ford Motor Company, and president of the World Bank. Haas became the chairman of Levi Strauss. Goodwin became the chief urologist at the UCLA School of Medicine, and Johnson became an eminent trial lawyer in San Francisco.

Hewitt, McNamara, and Haas went east to attend Harvard University's Graduate School of Business Administration. After the first year, money became an

issue, and Hewitt returned to San Francisco to work. He was with the accounting department of Standard Oil from 1938 to 1939 and the Texas Co. (Texaco) from 1939 to 1940. In 1940, he began writing advertising copy for Roos Brothers, a San Francisco menswear firm.

Soon after Pearl Harbor was attacked on December 7, 1941, Hewitt enlisted in the U.S. Navy. He served in the Pacific theater from 1942 through 1946, and was commissioned as an ensign on the USS California. In the next five years, he earned nine battle stars and four medals, including the Presidential Unit Citation Medal. He left the service with the rank of lieutenant commander.

Hewitt gained another lifelong friend in his cabinmate, Gabriel Hauge, during the war years. Hauge was the chairman of Manufacturers Hanover Bank from 1971 to 1979. He was also treasurer to the Council on Foreign Relations from 1964 to 1981 and a member of the Bilderberg Steering Committee.

After the war, Hewitt began working as a territory manager for Ford-Ferguson Tractors. He soon met Patricia "Tish" Deere Wiman, an outstanding young woman from Chicago whose energy and vision matched his own. A lifelong friend later said that Tish couldn't have hidden if she had wanted to. She was over six feet tall, had flaming red hair, black eyebrows, and dressed with flair. Her father, Charles, was the president of Deere & Company. Tish and Bill married in 1948. Bill was appointed as territory manager in the San Francisco area for John Deere Plow Company in 1948 and as general branch manager in 1950. As the branch manager, he was mentored by Benjamin Keator. Keator was a member of the Deere family.

According to author-historian Wayne G. Broehl Jr., young Hewitt brought an unusual sense of style to the operation and demonstrated coolness under pressure. As Broehl wrote, "Many years later, a *Fortune* magazine editor who interviewed branch personnel reached the following assessment of Hewitt, 'He conducted himself with grace . . . a good listener . . . did not throw his weight around . . . disarmingly cordial and acutely sensitive to others . . . a good leader.'"

Patricia (Tish) Wiman and Bill Hewitt made a glamorous couple in southern California after World War II. This 1947 photo was taken shortly after they became engaged. They were having dinner à deux at the Palace Hotel in San Francisco. *Anna Wolfe collection*

The young couple took a vacation to France in the late 1940s. Business journalists later said it played an important role in Hewitt's vision for the company he would lead. The trip left him surprised to find farmers in a modern country still plowing with horses and oxen. He saw an untapped market for John Deere farm machinery beyond the oceans.

Hewitt was elected to the board of directors of Deere & Company in 1950, and was coached by board colleagues in company history, tradition, and policy. Four years later, president Charles Wiman became terminally ill and he turned to Keator for advice. In a letter to Wiman dated May 26, 1954, Keator wrote a prescription for the position of president. He had weighed his fellow directors and concluded that Wiman's son-in-law, Hewitt, had the most important qualities, though not the experience.

I have felt, of course, that Bill has been pushed along pretty fast but I must say he seems to be able to absorb it in his stride without its affecting his personality or his association with others. He has had enough Branch House experience to know that problem pretty well and he is alert to the requirements in the field. He has you, he is smart, catches on to things quick and has vision, enthusiasm and courage to get things done . . . he has had a broad field to operate in and had the authority to act. This is important if he is to be able to face up to the task . . . To be sure, he has had little or no experience in factory operation, which is of course very important in our Company, but couldn't someone be appointed to his staff who could tutor him in these decisions? . . . Bill is a very dignified and personable young man and you would go a long way before you would find his equal for that job.

Hewitt soon was appointed to the new post of executive vice president. He chaired most board meetings through the fall and winter of 1954 and 1955, receiving major support from the board. On May 24, 1955, twelve days after his father-in-law died, Hewitt was elected the sixth president of Deere & Company. A very busy summer and fall lay ahead for the decorated ex-naval commander. Aside from business, Tish Hewitt gave birth in July to twin daughters, Adrienne and Anna Deere Hewitt—the sixth generation.

New Attitude

As Hewitt began to chart the course for Deere, the historic leader in the agricultural industry was International Harvester. Deere sales were second, closely followed by Ford. Massey-Ferguson and Allis-Chalmers were substantial, but smaller, competitors. The Oliver, Case, and Minneapolis-Moline companies also had many faithful tractor customers.

Over the next decade, Hewitt instituted changes that redefined the industry role of Deere & Company. The company already had a long history of making high-quality products tooled to the needs of farmers, and it was profitable for shareholders. It had a casual, honest, Midwestern, small-town attitude. It was all-American and modest.

Developers could hardly keep up with demand for new housing during the postwar Baby Boom. In this undated photo, a 1950 two-cylinder Model M John Deere crawler works at landscaping in one of the new subdivisions. *Deere & Company*

Former officers like Hewitt, discharged after World War II, were a strong new resource for the old company. Gerry Mortensen, a retired research engineer, recalled that Deere acquired a large number of personnel who were graduates of the Navy V-12 officer-training program. They recognized the immense value of some high-tech wartime military procurement and manufacturing techniques, such as statistical quality control. These techniques would prove invaluable as Hewitt later worked to encourage new product development and efficiency in production.

Deere manufacturing adopted the statistical quality control program quickly and completely. The results were marketed as the John Deere Quality Farm Equipment program. Competitors laughed at the statistical quality control program for years, while versions of it were being learned and adopted eagerly in Japan. Two decades later, Mortensen recalled, they were scrambling to import the revolutionary manufacturing breakthrough while Deere

was enjoying the benefits. Deere, in the 1970s, possessed a highly sophisticated manufacturing group, more modern manufacturing equipment than the industry norm, and top-level personnel who understood how to best use these advantages.

Although Hewitt was among the youngest members of the board and had relatively little company experience, he was also a family member, and therefore had their basic goodwill for support. He represented the next generation to most board members. For his part, Hewitt saw entrenched attitudes that would resist change and an old organizational structure. He also saw a stage already set in 1953 for replacing the two-cylinder tractor, Deere's primary product. Researchers had begun to design multi-cylinder engines with supporting transmissions and hydraulics.

Within weeks of becoming the new commander, Hewitt launched several elements of change that he wanted to see. During the fall of 1955, he assembled a set of objectives for the company. He believed that Deere could and should become number one in the North American farm machinery industry. This would include the areas of sales, profit ratios, product quality, new design, safety of operations, and excellence in relations with employees, dealers, stockholders, and the public. Broehl commented:

This was a herculean set of goals. Had one queried Deere's then top management, they most likely would have said, "We're a good number two to International Harvester—let IH innovate." Hewitt challenged this thesis within the first two weeks of his election. In a speech to all the branch house managers he said: "Yesterday I received a friendly letter from one of our most active competitors, and emblazoned across the bottom of the stationery in splashy red ink was imprinted: 'We are not aiming to be runner-up.' Believe me, we in Deere & Company are not aiming to be runner-up either—we are aiming to be first in all our business activities." The gauntlet had been laid down: "Pass International Harvester."

This was a new attitude for the old family-run corporation, but it was backed with determination. Deere & Company reinvented itself in the next decade—in corporate structure; in overseas manufacturing; in a new headquarters; in product development, promotion, and support; and in its public image.

New Structure

An organizational shakeup began in the first month of Hewitt's leadership. Meanwhile, the John Deere Tractor Company in Waterloo, Iowa, was launching the first fully integrated, factory-designed power steering system for farm tractors, and was preparing to introduce its largest two-cylinder tractor, the Model 80. The newer factory along the Mississippi River, the John Deere Dubuque Tractor Works of Deere Manufacturing Company, was gearing up to introduce its new 20 Series tractors in a few months.

Already, Hewitt had received a confidential proposal for a merger with Massey-Harris-Ferguson. (Just two years earlier, the worldwide operation of Massey-Harris Company had merged with the Harry Ferguson Company.) Hewitt was willing to discuss the proposition of a merger but startled his directors with a request of his own. He proposed that they hire an outside management company, Booz Allen Hamilton (BAH), to study the organizational structure and executive compensation at Deere & Company. Suggesting they conduct a study of the company's organizational structure was not a new idea, but past evaluations had been done in-house. Most directors felt the company was too idiosyncratic for outsiders to understand. With some reluctance, they supported Hewitt's request. The study was delivered in January 1956.

The BAH study recommended, and led to, major adjustments. A reorganization was announced in April 1956. The corporate structure was altered to give the home office more authority to coordinate the work of the branches and factories; executive compensation was brought in-line with the rest of the industry; and key older executives were eased into new advisory positions to make way for a new generation of Deere managers.

"There was little doubt in anyone's mind after this critical meeting, just eleven months after Hewitt had assumed the office of chief executive, about 'who was in charge,'" says Wayne G. Broehl Jr. in his book *John Deere's Company*. "Hewitt had put his mark on the management structure decisively, and with sensitivity and aplomb."

Shortly after, a name change took place. The Waterloo factory became John Deere Waterloo Tractor Works on June 1, 1956. The Dubuque factory became John Deere Tractor Works of Deere & Company on August 1, 1958, followed by a change to a simplified name, John Deere Dubuque Works, on November 1, 1974. Further name changes followed at Waterloo in 1975 and 1980.

New Headquarters

While Deere customers worked fields in the summer and fall of 1955, the newly elected president was searching for

Scouting for a site in 1956, Bill Hewitt and architect Eero Saarinen used a truck with an elevating platform to get an over-the-top view of potential places for the new administrative center. *Anna Wolfe collection*

Bill, Sandy, and Tish Hewitt at the construction site for the new Deere Administrative Center in Moline in 1961. The occasion was to observe the raising of the first steel column into place. *Anna Wolfe collection*

Forum, pointing out the new Kresge Auditorium at the Massachusetts Institute of Technology, and another magazine showing the new General Motors Technical Center. Both buildings, Dreyfuss suggested, would be good models to emulate and both designs were by the internationally acclaimed Finnish American architect Eero Saarinen.

Hewitt invited Saarinen to Moline. The two visited four potential sites in August 1956. They borrowed a utility truck with a telescoping tower so they could look over the tops of smaller trees. They eventually settled on a site that consisted of four farms with about 720 acres total, roughly three miles south and four miles east of the downtown Moline office and about two miles west of Bill and Tish Hewitt's new home. About half the site was perched on hilly bluffs overlooking the Rock River. Saarinen sent a submission for developing the site soon after. Hewitt wasn't satisfied with the first submission, but he accepted the second. The plans were approved by the board in January 1957.

Seven years later, the new John Deere Administrative Center on John Deere Road was complete and open for business. The new complex received international acclaim and won dozens of architectural awards. It came to be recognized as one of the finest corporate administrative centers in the world. Deere & Company's headquarters even had a new address: 1 John Deere Place (replacing 1325 Third Avenue), Moline, IL.

Hewitt's daughter, Anna Hewitt Wolfe, recalled in 2009 in personal correspondence that: "My father was particularly proud of the fact that the building received an award 25 years later from—I believe it was the American Institute of Architects. This showed that the building was still outstanding 25 years later, and that he certainly had made a good decision back in the 50s."

According to Broehl, the plan was expensive, perhaps extravagant. Management objected to the cost. Some feared the project might create a negative image in the minds of farm customers. Hewitt focused the directors on the choice of architect rather than the building. He described other buildings around the world by Saarinen.

a site for a new corporate headquarters. The company was 120 years old. It had started in a small Midwestern town and stayed there, only growing in a conservative fashion to meet its basic functional needs. It had many buildings for offices and factories, but they were the products of company engineers rather than design architects. Hewitt envisioned Deere's leadership in the farm machinery industry, and he wanted the corporate offices to have a look and feel that created a national presence in architectural style.

Hewitt contacted two people to launch the project: Robert McNamara and Henry Dreyfuss. McNamara had supervised construction of a new administrative building for Ford in Detroit. At Hewitt's request, McNamara sent his friend a packet with background information on two dozen architects. New York City industrial designer Henry Dreyfuss had been contributing to the design of John Deere tractors since 1937. When asked for ideas, he sent along the July 1955 edition of *Architectural*

Then Hewitt posed a question to the board that seems in retrospect to epitomize more than any other single statement or act the direction in which he wanted to take the company. His query: "Is he too fancy for us?" His answer, "No."

Hewitt was asking the board to enlarge its concept of the company—to see itself as a major, nationally important entity, on its way to becoming the pre-eminent firm in the industry. In effect a statement was to be made by the company, one that Hewitt hoped would reverberate through the organization at all levels and out among the dealers, the customers and the general public.

Shortly after the new headquarters for Deere & Company opened in Moline, Illinois, former President Dwight D. Eisenhower came to visit and have lunch with Bill and Tish Hewitt. *Anna Wolfe collection*

Hewitt firmly believed the new center would have a positive impact on the company's image among its farm customers. Equally important, he believed leadership would help the company attract and retain high-quality employees. In retrospect, building the new center also contributed to winning the allegiance of management old guard to Hewitt's vision of the company's goals and standards.

When completed, the seven-story, 330-foot-long main administration building stretched across a gentle valley and faced a landscaped approach. The approach included two artificial lakes. Later, Henry Moore's *Hill Arches* was added on an island in the upper pool. The building contained dining facilities for 1,000 employees. The exterior was a mixture of steel and glass, with beams, columns, and balconies reminiscent of Japanese construction. The iron building would become a rusty palace due to the weathering of the steel, a look complimented by bronze plate glass and ideally suited to the home of a farm machinery manufacturer. As recently as 2000, the *Source Book of American Architecture* described it as "a stupendous building with an integrity not likely to be seen again soon."

New Markets

Hewitt's election as president also set the stage for an expansion of Deere & Company into manufacturing beyond the borders of the United States and Canada. Before that time, unlike major competitors, Deere did not operate outside of the United States and Canada. International Harvester had been operating overseas for decades. Canada's Massey-Harris, the largest agricultural equipment maker in the British Empire, had been active in the U.S. market since acquiring a New York state company (Johnson Harvester) in 1910. J. I. Case Company, by 1929, had expanded to Australia, Mexico, Sweden, and other countries. Ford had ended tractor production in North America in 1928 but continued building tractors in Ireland and later in England. Deere stayed "home" while the others built abroad. In fact, as late as 1952, Charles Wiman's proposal for manufacturing in Scotland had been aborted; the board had not been willing to commit.

Hewitt saw Deere's lack of foreign operations as a major handicap in his quest to lead the farm machinery industry. In early 1956, days after the collapse of negotiations for a merger with Massey-Harris-Ferguson, Hewitt started looking for manufacturing opportunities abroad. He sent one team to Europe and another to Mexico on scouting missions. Both teams returned with enthusiastic reports. Encouraged by the reports and the new leader, Deere's board agreed to fund expansions into Germany and Mexico.

An old, struggling German tractor manufacturer was given new life in 1956 when controlling interest was purchased by an up-and-coming American company, John Deere. This 36-horsepower Lanz semi-diesel tractor was one of several dated models still being built in 1955. With new paint, it has migrated into the public collection of the Manitoba Agricultural Museum at Austin, Manitoba.

In late July 1956, Deere directors approved the purchase of 51 percent ownership in Lanz, a small, struggling, but well-established German farm equipment manufacturer. Deere now owned the Lanz "Bulldog" tractor factory at Mannheim and the factory at Zweibrucken, where harvesting equipment was manufactured.

In a second international purchase a few months later, Deere invested nearly two million dollars in Mexico. It purchased land at Monterrey and began construction of an assembly plant for John Deere tractors and some implements.

In September 1957, the board authorized construction of a third project, a tractor assembly plant in Rosario, Argentina. It would have enough capacity to produce 3,000 tractors per year for Argentinian farmers.

In 1959, a fourth international project began almost by chance. Deere learned of a tractor development effort started by three small French agricultural machinery makers. They had formed a company, Compagnie Continentale de Motoculture (CCM), and allowed Deere to purchase a 51 percent controlling interest. The first green-and-yellow tractors coming out of this new facility in 1965 were destined for German farmers. They operated with three- and four-cylinder diesel engines.

Hewitt started a fifth overseas project in 1962. Deere purchased 75 percent of the stock in a South African implement manufacturer, which enabled the company to introduce its own tillage products and import and assemble tractors at Nigel, a town east of Johannesburg.

A window in Asia seemed to open in 1963 when Deere entered a licensing agreement with Hitachi, one of Japan's largest manufacturers. Hitachi agreed to build certain Deere tractors, and equipment, for farmers in Japan, South Korea, and Okinawa. In turn, Deere would sell some Hitachi machines and provide technical assistance. The plans, however, gradually came apart. Hitachi

This hi-crop version of the John Deere two-cylinder 730 wasn't built in Waterloo. It was built with equipment from the Waterloo factory that was shipped and reassembled at the Deere factory in Rosario, Argentina, in the 1960s. It was imported for the Keller Tractor Collection at Brillion, Wisconsin, and has been fully restored. *Bruce Keller*

lost interest when pressure from Japan's government eased off. The two companies agreed to end the licensing agreement in 1970.

Generally, the foreign operations were not profitable during Hewitt's first 10 to 15 years of leadership. There were substantial problems learning what foreign customers wanted and dealing with foreign government regulations. Hewitt was patient; in his second decade at the helm, overseas profits finally began to appear.

Product Development: TOP SECRET

When Hewitt became president in 1955, Deere & Company was already building a long line of machines and implements for farming, and had initiated a monumental project to develop multi-cylinder tractors. The two-cylinder tractor was, by far, the most important product built by Deere, although the company was also building equipment for tillage, planting, haying, harvesting, and farm chores like loading, hauling, lifting, hammering, and shaping.

Starting in 1913, the Waterloo factory had manufactured more than 1.1 million two-cylinder tractors by the time Hewitt became leader. The newer factory at Dubuque had built approximately 137,000 tractors, and a third factory at Moline had built more than 27,000 small two-cylinder tractors between 1936 and 1946.

In an exhibit area, a new John Deere 4020 row-crop tractor forms the backdrop for tractor talk between dignitaries. A notation on the back of this photo reads: "Eisenhower/Papa/Maman talk tractors and farming at Deere & Co., May 1965." *Anna Wolfe collection*

The two-cylinder engine had uniquely fulfilled farm demands, and provided profitable returns. The Johnny Popper was easy to sell, cheap to operate, easy to understand, and farmer-repairable. It had become a point of pride that John Deere two-cylinder engines were performing with, and outperforming, tractors with more cylinders. By 1937, in fact, Deere put out an official statement on the subject. Standing behind the two-cylinder engine, L. A. Rowland, vice president and general manager, attached his name to this proclamation:

A FOUR-CYLINDER ENGINE—NOT ON YOUR LIFE.

Some of our dealer and tractor owner friends occasionally tell us they have heard that John Deere is coming out with a four-cylinder engine in all John Deere Tractors. Dealers and others who are well informed know that this rumor is false.

We want to assure those who are not so well informed that we are NOT coming out with a four-cylinder engine. THE JOHN DEERE TWO-CYLINDER ENGINE HAS BEEN SO OUTSTANDINGLY SUCCESSFUL THAT THERE IS NO THOUGHT OF A CHANGE.

Calls for larger engines increased after World War II. Thousands of young farm men were now familiar with the four- and six-cylinder engines, both gas and diesel, that were used in military equipment like tanks and jeeps. On their own initiative, they began adapting larger engines to existing tractors. Going partway to respond to new expectations for farming, Deere introduced its first diesel engine on the new Model R two-cylinder tractor in 1949.

Early in 1953, after a change of directors, a high-level decision was made to do the unthinkable—develop a Deere multi-cylinder engine for tractors. Directors had

An element that set Deere & Company apart in the 1960s and 1970s was a series of eloquent statements on the company's philosophy, a type of "constitution." It was called the Green Bulletin at Deere, and went by different names in other leading corporations. The Green Bulletin was vague and general; it didn't command or control, but it did present the "great ideas" that governed the company. The purpose, said Deere Executive Vice President Tom Gildehaus, was to convey "the cultural expectations" of the company, a stable core of moral values.

"There is a broad sense of humanity in the series of Green Bulletins. They convey to managers that all decisions must reflect a balance between business and human needs. But, most important, the message is that we must lean to the human side when a difficult choice must be made."

Hewitt issued the first John Deere Green Bulletin in the summer of 1964 at the opening of the new administrative center. Additions were distributed by Hewitt in 1966 and in 1975. They have been re-issued by successors since 1982. As updated and reissued in 2004, there are twelve points to the Green Bulletins. The first is a vision statement: "Our vision, our aspiration, is to distinctively serve customers—those linked to the land—with a business as great as our products. Such a business consistently delivers greater value to all who have an interest in our success. Achieving our vision requires exceptional operating performance, disciplined Shareholder Value Added (SVA) growth, and aligned, high-performance teamwork."

Other points included a set of four core values (integrity, quality, innovation, and commitment), commitment to employees, to customers, to shareholders, global business and diversity, optimized teamwork, business ethics and compliance, the environment, health and safety, public responsibility, "performance that endures," and protection of the John Deere brand.

Hewitt declared the most important values, for continuing success both as an organization and as individuals, were integrity and merit. He wrote that integrity and merit consist of these elements:

COURTESY in every word.
HONESTY in every transaction.
DIGNITY in every personal act.
PROGRESSIVENESS in every thought.
CONSTRUCTIVENESS in every criticism.
QUALITY in every piece of work produced.

accepted that farmers were pressing for refinements like better steering control, independent power takeoff, more hydraulic power, more horsepower for heavier jobs, and better transmissions.

The effort and investment required to meet that demand was staggering. The production machines in Waterloo's factory were aging. They were less accurate than 15 years earlier, and had more downtime for adjustments and repairs. Engineers recognized that their primary tractor factory would itself need an overhaul, as well as new machines, to build tractors and engines to new tolerances.

Directors recognized that the whole tractor factory would need to shut down for months of retooling, and a major support effort would be required to prepare the supporting infrastructure. New tractor models required new operating and service manuals, training for service people, the manufacture and stocking of new parts, and promotional materials. All the while, the product had to remain under a tight cloak of secrecy so that competitors would be caught by surprise and sales staff could continue to sell the existing stock of two-cylinder tractors.

Merle Miller, a senior division engineer at Waterloo, worked with Deere from 1946 through 1984 on projects

The largest and last of the John Deere two-cylinder tractors, the 830 is prized by many collectors today. The big engine chugged along at 1,125 rpm and was rated for about 70 horsepower on the drawbar. Shipping weight was nearly 12,000 pounds. This restored model is in a private collection in Manitoba.

such as the 8010, 1020, 2020, rollover protection, enclosed cabs, and later tractors. In his book, *Designing the New Generation John Deere Tractors*, Miller related the history of the New Generation tractor design from his perspective, writing in April 1953 that 20 Waterloo engineers and design men were pulled off regular duties and appointed to start developing an upright, multiple-cylinder engine. The research group set up shop in a rented former grocery store dubbed "the meat market." It was top secret from the start. The group hoped to deliver the new models

to dealers within five years, but it took seven. Another research group, headed by Wayne Worthington, converted existing products to learn how much horsepower could be used efficiently in a two-wheel-drive, rubber-tired farm tractor.

In one conversion, they installed a four-cylinder 471 Detroit Diesel engine, a torque converter, and instrumentation into a Case LAI (Industrial) tractor with 18x26-inch tires. Designated as RTX547, the test tractor spent two seasons in Montana and one winter at the

Texas test farm. It was tested with a variety of field tools. Worthington's team concluded, in 1955, that at least 100 horsepower could be utilized and new designs should be expandable to that level.

According to Broehl, the design effort lost a few months by trying to accommodate marketing managers in branch offices who wanted something exclusive in the engine design to help sell the new product. To keep the peace, the design team tried to create a V-4 or V-6 tractor engine. "The V-design engine had worked well on the automobile, but the idea was flawed for tractors—the narrowness required of the row-crop tractor would not allow the requisite juxtaposition of the pistons. After months of work on the V, the notion was abandoned. The engineers were forced to turn to vertical positioning."

Concerning these same critical months, Miller wrote:

Several design studies showed the problems of increasing the engine power in the two-cylinder tractors . . . Increasing the displacement or adding cylinders on the horizontal two-cylinder engines meant the width of the tractor would become larger across the cylinders. This was unacceptable on row-crop tractors . . . Increasing engine speed was inherently difficult on the two-cylinder engines because of vibration . . . With the two-cylinder engine limited by width restrictions, other options became possible for engine arrangement. A new tractor design would require new factory equipment, but allowed for considerable latitude in concept.

The engineering team was enthusiastic, but needed a better place to work than the rented grocery store. The summer that Hewitt became president, directors committed to building the new Product Engineering Center. Land near Waterloo was purchased, and a building shell was erected before winter. Insulation, heating, and interior plumbing were installed after the offices were in service. The cold, rough first winter didn't stop the Deere design engineers. They were designing engines, and building a revolutionary new line of tractors from the ground up.

The Product Engineering Center official opening came in 1956. It was state-of-the-art, and included one of the first room-size computers. The world also noticed that Deere machinery had a new trademark that year. It featured only the silhouetted deer and the words "John Deere" on a black background.

Closing the Two-Cylinder Era

While a small army of employees gradually geared up to produce a new generation of tractors, engineers and designers supplied innovations that helped maintain two-cylinder sales. The whole line of Deere tractors from the two factories consisted of about 30 model configurations in six power sizes. A number designation (40, 50, 60, 70, 80) had replaced alphabet names for models about the same time the multi-cylinder engine was taking shape on drawing boards. These were upgraded to the 20 Series in 1956. Two years later, the 30 Series arrived at dealerships.

The last power boost for Deere's two-cylinder tractor engines was provided in the 20 Series tractor fleet. It had approximately 20 percent more power than the fleet it replaced. Generally, these tractors were also highly successful for Deere and gained market share for the company. They had many refinements in hydraulics, fuel systems, and engine design. They were more fuel-efficient and could pull equipment a bit faster, or could pull slightly wider equipment. They set fuel economy records in 10-hour runs at the Nebraska Tractor Test Laboratory. They were more comfortable for operators, and the styling was more attractive.

When the 30 Series came along in 1958, it had not received much engineering attention but did have a number of attractive cosmetic upgrades. New fenders, for instance, added a much better safety level for operators. The new look helped keep dealerships busy. Internally, it provided a two-year window, some breathing space while Deere completed tooling up for its New Generation tractor lines.

The last and smallest John Deere two-cylinder diesel tractor from the 1950s was the 435. In the Nebraska test performed during September 1959, the tractor developed nearly 33 horsepower on the PTO. It was built in the Dubuque, Iowa, factory of John Deere, but it featured a General Motors diesel engine governed at 1,850 rpm. Photographed in the Gordon Gilchrist Tractor Collection, Wainwright, Alberta.

Deere's largest, most powerful two-cylinder tractor was the Model 820 and its successor, the Model 830. The Model 820, equipped with Powr-Trol and Power Takeoff, weighed 8,150 pounds. It also had a touch of style—bright yellow hood side panels.

Custom Powr-Trol was big news when it was introduced on the Model 820 in July 1956. It came out on the other 20 Series tractors in the next few months. Custom Powr-Trol was completely independent of the transmission clutch and controlled three separate hydraulic circuits. Draft Control forwarded implement loads

through the top link and, by pushing on the hydraulic control valve, the operator actually controlled an implement's working depth. These tractors also had two rear hydraulic outlets beside the PTO and an emergency disconnect system.

Hydraulic power control, using closed-center pumps, was state-of-the-art at the time, but it had serious drawbacks that PEC engineers resolved for the New Generation tractors.

New Generation tractors had a unique and innovative "closed-center" hydraulic system. It featured significantly higher working pressure for increased workload, zero flow at idle for significantly reduced heat load and power waste, zero leakage to stop cylinder drift, poppet valves to eliminate sticking, and a unique piston pump that only Deere was equipped to manufacture.

Later, using this system, Deere production attained high-volume, low-cost performance when it introduced multiple lines of industrial equipment (loaders, graders, and backhoes). Competitors addressed the issue too, but their hydraulics were more complex and less cost-efficient to produce, leaving Deere with a significant marketing advantage.

Change to Come

Deere's Industrial Marketing Division was established in 1957. Based on production from the Dubuque Works, that factory was drawn deeper and deeper into designing and manufacturing commercial construction equipment.

In 1958, the Dubuque factory began producing a publication about industrial equipment for dealers, and introduced industrial yellow Model 440 wheel and crawler tractors for the first time. Commercial customers could order an all-hydraulic bulldozer. The 440 Series offered heavier hoods and grilles, a new single-stick control for both steering and shifting, and a 10 percent boost in power.

A hint of engine changes to come was revealed in 1959. Deere & Company introduced two tractor models using a diesel engine from another manufacturer; one of these had six cylinders.

The Model 435 was a small tractor from the Dubuque factory. It had a 106.1-cubic-inch two-cylinder General Motors (Detroit Diesel) two-cycle diesel engine. The Nebraska Test, with the first 1,000 rpm PTO test, rated it for 32.91 horsepower. The Model 435 was also the first tractor to use the new three-point hitch dimensions standardized by the American Society of Agricultural Engineers.

The Model 8010 was a huge tractor. It came out of Waterloo, and was introduced at the John Deere Field Day in Marshalltown, Iowa, in 1959. It was a mammoth four-wheel-drive articulated tractor with power steering.

It weighed 24,860 pounds when the tires were full of ballast, more than twice the weight of the Model 830. It stood eight feet tall at the top of the steering wheel, was eight feet wide, and measured more than 19 feet long. It was powered by a 425-cubic-inch General Motors 671E two-cycle six-cylinder diesel engine with an estimated 215 horsepower. The transmission had nine forward speeds ranging from 2 to 18 mph. A 24-volt electrical system provided fast starting in any weather, and air brakes made it easy to stop.

The Model 8010 also was about ten years ahead of the market: it was just too big. It could pull a 31-foot harrow and plow at up to 7 mph. The three-point

A rare operator's manual for a rare tractor, the John Deere 8020. Only 100 of these 25,000-pound tractors were built. They offered an articulating body and four-wheel drive. Several companies were experimenting with these concepts in the 1950s, but it took another ten years for the tractor and the market to really develop.

hitch easily lifted an eight-bottom plow. *Two-Cylinder Magazine,* in its January-February 1997 issue, reported that only 100 Model 8010 tractors were built, and one stayed at the factory. All 99 others were recalled and modified, becoming known as Model 8020.

In a day's work, the 8020 could field cultivate 260 acres, plow an easy 50 acres with an eight-bottom integral plow, double-disk 185 acres with a 31-foot harrow, and perform other remarkable feats. It took dealers seven years to sell the last of the 8020s.

The tractor-plow combination was demonstrated at Deere's largest volume dealership, Grossenburg Implement near Winner, South Dakota. Authors of the company history, *Grossenburg's Fifty Years with John Deere 1937–1987,* later commented, "the time had not arrived for the use of four-wheel drive tractors in the Rosebud area."

Late 1950s Products

Deere wasn't the only company presenting new products for farming in the late 1950s. Tractor-makers engaged in an all-out horsepower race. Fewer people had more land to farm, and field speeds needed to increase. The horse-and-plow rate of about 2.5 mph shot up to more than 4 mph. Tractor transmissions were strengthened to handle the greater engine power.

International Harvester, as early as 1954, introduced a torque amplifier. It allowed a standard five-speed transmission to act like a ten-speed. In that mode, pulling ability increased nearly 50 percent. Three years later, J. I. Case replaced the standard clutch with a torque converter. Broehl wrote, "When it was locked in, the tractor had an exceptionally good ability to start with a heavy load and, after gaining requisite operating speed, the torque converter could be locked out. The converter could be reengaged at any time to provide as much as a 100 percent increase in pull over that of direct drive, thus allowing the tractor to pass more easily through rough spots in the field." The torque converter had a problem, however. Mortensen recalled that farmers did not like "the windmill effect" on the engine and the sound, which was "like a Buick Dynaflow."

A change of fuels was also underway. Distillates had been popular for many years, but were phased out in the late 1950s. The last distillate tractor tested by the Nebraska Tractor Testing Laboratory was a John Deere 720 all-fuel tractor, in 1956. Diesel fuel now was readily available and accepted as a better choice.

On other fronts, Deere's successful harvesting products held their own or had minor improvements in the 1950s. Improvements and minor changes were made throughout the company's products, including cotton pickers and plowing, cultivating, and haying equipment. Deere introduced the two-row Model 10 corn head in 1954 for the new Model 45 combine. It was the first corn head in the industry. A history of the Grossenburg dealership in South Dakota states: "Since then, John Deere has been the unchallenged leader for corn-head and row-crop attachments. This is attested to, by the fact that all major brand combines have attaching brackets to adapt John Deere corn-heads and row-crop attachments to their own machines."

A John Deere Day for dealers to show off new equipment had been introduced at the 100th anniversary in 1937. The 1957 John Deere Days, according to Grossenburg, drew 1,700,000 farmers nationwide—about one out of every four farmers in the United States. The 1957 innovations included six-row planters and cultivators. These increased planting and cultivating capacity for row-crop farmers in corn- and cotton-producing areas by 50 percent.

A 1958 introduction was the revolutionary John Deere 214-T bale ejector from the Ottumwa Works in Iowa. It helped maintain Deere's leadership in haying equipment, as an improvement on the John Deere 14-T automatic twine-tie baler that had been introduced four years earlier.

Deere completely redesigned its trademark magazine for customers in 1958. Published since 1895, *The Furrow* was already the oldest company-based farm magazine and was well read by a wide audience. From the start, it was a unique magazine. *The Furrow* had pioneered the concept of content marketing by offering straight, relevant news

This John Deere 14-T square baler still produces bales each summer on a small Manitoba farm. When it was introduced in 1954, it was the first automatic twine-tying pickup baler and an immediate success. A larger unit, the 214-T, was introduced in 1957. Parts are still available through dealers.

to customers without pushing John Deere products. Aside from advertising, the publication offered news on farm trends, developments in crop science, and innovations like weed control and soil conservation. It was worth the time to read because the content was different from other publications, and it wasn't available by subscription. The only way to get it was by a personal request to the local John Deere dealer.

In 1958, at Hewitt's direction, *The Furrow* became a template for modern corporate marketing. It became the first full-color, high-quality magazine for customers produced by a corporation.

With the 1958 edition, *The Furrow* had smaller pages but more of them, and contained full-color photos to accompany the stories. This made *The Furrow* a pioneer among corporate magazines. To achieve this, a Chicago printing company was hired and new printing equipment was specifically designed and built for *The Furrow*. The

new design was on a modern 8½x11-inch format and printed on high-quality glossy paper. Regional editions had up to 48 pages and were published at a rate of six to eight issues yearly. A Spanish-language edition was launched in 1959 for farmers in Argentina. New editions were launched for Spain in 1960 and for German-speaking farmers in 1964.

Another important introduction in 1958 was the John Deere Credit Company. It was created to provide funding for John Deere retail financing programs. John Deere Credit, as it is known today, operates with more than 1,700 employees in 16 countries and is among the largest finance companies in the United States. It has two major goals: supporting equipment sales by providing financing to customers and providing a variety of financial services to rural communities.

In the late 1950s, Deere's dealership network gradually was trimmed and shaped into a small, sophisticated

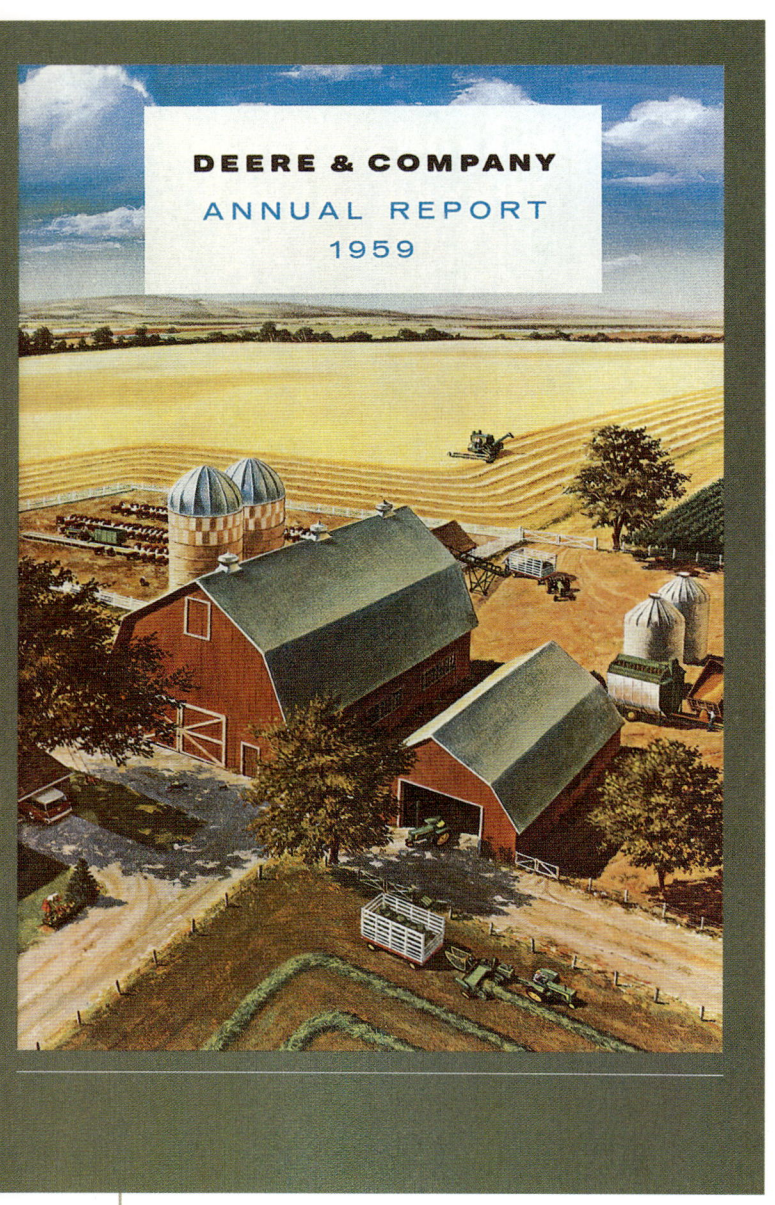

Deere & Company Annual Report, 1959.

But for the first time, in 1958, Deere's sales of farm equipment in the United States topped International Harvester. That year, while Deere's new 30 Series tractors were generating loyalty and respect, International Harvester was dealing with repercussions from its hurried introduction of the Farmall 560.

The flagship 560 tractor became notorious. It wasn't ready for heavy jobs, and it soon was subject to a product recall nightmare. International had to offer an entirely new rear end for the tractor by mid-1959, and sold only about 8,000 in the first two years. There wasn't time to recover market position before Deere introduced their New Generation tractors in 1960. The international giant was in trouble. Deere wasn't yet actively manufacturing outside of North America, but it had solid leadership, a retooled tractor facility at Waterloo, and a modern one in Dubuque. In addition, the company had avoided major labor disruptions and had stayed profitable with very successful lines of products.

Author Barbara Marsh described the transition of power in a book called *A Corporate Tragedy: The Agony of International Harvester Company*:

> *Dealers resented Harvester's near-insistence that they erect large new facilities to accommodate the company's postwar sales pitch . . . Harvester's preoccupation with the growth of its truck and refrigeration business hurt its image in farm machinery as the company forced its farm equipment salesmen to peddle its other lines as well . . . Harvester also relinquished its market leadership because it failed to meet the American farmer's postwar demand for bigger equipment . . . Harvester's historic success with the Farmall tractor has fostered a peculiar conceit among Harvester's sales-oriented executives who assumed the company's leadership would last forever . . .*

The author went on to talk about the reasons that Deere & Company was able to overtake International Harvester:

system. Hewitt stepped up training for territory managers so they could be better advisors. They helped tackle problems like ordering parts, stocking, and inventory turnover. The company placed parts specialists and technicians in the territories to help managers. Dealers and staff also attended business improvement workshops.

Meanwhile, for a variety of reasons, the giant of the machinery industry was stumbling badly. International Harvester had troubles on several fronts. In the 1930s it had been up to eight times the size of Deere & Company.

International Harvester was the world leader in farm machinery sales in the first half of the twentieth century. The company suffered heavy losses in market share against Deere in the 1950s. In an effort to regain ground, it launched five new models in 1958. The largest was the 560, a six-cylinder diesel meant to be directly competitive with Deere's 830. This unit was assembled from two of the 560 tractors by a Manitoba enthusiast, near Neepawa.

Deere had become unbeatable by giving its winning dealership network a top-notch line of big equipment, tailored to the growing needs of big farmers. In the 1950s, Deere obliged the big farmer by introducing six- and eight-row planters, the first combine for dual harvesting of corn and wheat, and a new hay baler enabling one man to do a two-man job. Deere, riding on its superior financial strength, also offered easier credit terms to farmers. In 1957, the firm instructed its dealers to take any farm notes that were "sound," a move that attracted customers. Between the beginning of 1956 and the end of 1958, Deere's retail receivables more than doubled to $132 million.

By the end of the decade, Hewitt was ready to introduce tractors that represented a new generation of power. They would seal Deere & Company's grasp on first place in industry leadership.

The 1010 and 2010 were the two smallest New Generation tractors introduced in 1960. They were built in the Dubuque factory. This 2010 had the new Syncro-Range transmission and new hydraulics. It was fully restored by Tony Gerber of Millbank, Ontario, and is in a private Ontario collection. *Tony Gerber*

THE NEW GENERATION OF POWER:
1960–1963

D-Day in Dallas

On August 30, 1960, Deere & Company created a new piece of the American Dream when it unveiled the "New Generation of Power." Judging from the scale of the festivities in Dallas, Texas, Bill Hewitt was pleased with the product that was being unveiled. He treated the event with the showmanship one might expect of an ex-naval commander who took part in the celebration of victory at V-J Day in the Pacific, 15 years and 15 days earlier. The rollout of John Deere's new flagship line of machines was celebrated with nearly as much grandeur and pomp as a presidential inauguration.

The transformation from two-cylinder to four- and six-cylinder tractors had been a monumental task. Nearly every part was newly designed for these tractors, and these new parts required extensive tooling and testing. The Waterloo factory had been shut down, disassembled, cleaned out, and rebuilt with new machines in new assembly line configurations. Once the remodeled factory was open, totally new tractors began to be assembled. Dealers had to be stocked with parts, and much more work was required as well.

The event was unprecedented in Deere history, but a debut two years earlier by the long-time leader in tractor manufacturing, International Harvester, may have

provided inspiration and lessons. On July 15, 1958, International Harvester introduced what it billed as its "New World of Power" at a 65-acre site on the International Harvester testing farm in Hinsdale, Illinois.

John Deere New Generation tractors are attractively styled, reflecting changing tastes in the 1960s. This air cleaner and exhaust are part of a 4010 owned by Brent Gordon of Souris, Manitoba.

An estimated 12,000 International Harvester dealers from North America and 25 foreign countries watched the parade. The display included 76 Farmall tractors with 110 implements. Another 250 tractors and 260 implements were displayed or demonstrated. It was the biggest show ever for the world's leading farm equipment manufacturer. The six-cylinder, 62-horsepower Farmall 560 led the show. The new machine was touted as the "world's most powerful row-crop tractor," and it could plow 30 acres a day when hitched with a five-bottom plow and cultivate up to 100 acres per day using a six-row cultivator. The tractor featured silky-smooth power steering, a very comfortable operator's seat with a backrest, and excellent fuel economy.

Hewitt decided to surpass IH's impressive show in Dallas. The New World of Power was tainted by a bitter lesson in management for International Harvester. The 560 and its little brother, the 460, developed a serious transmission problem. Final drive failures began occurring in the first 50 to 100 hours of field operation. The bull gears, bull pinion, and brake shaft were revised in January 1959, followed by a modified bull gear hub assembly in March and a revised differential bevel gear design in April. Eventually, a $19 million field rework program was initiated. Tractors were shipped to centralized rework stations, drawing publicity that was deadly for sales.

The management practices of the two competitors were described in an important book about corporate management, *Vanguard Management: Redesigning the Corporate Future*, by James O'Toole.

> *The recipe for success at Deere has been integration, balance, consistency, and coherence. This stands in sharp contrast to the inconsistent, scattershot strategies of their prime competitors. It is useful to recall that Deere started far behind Cyrus McCormick's International Harvester, but IH squandered their lead with unwise, unrelated acquisitions (refrigerators, air conditioners), sloppy quality (they once placed a combustible gas tank on top of a red hot tractor engine), poor labor relations (they got tough with the UAW which cost them a 172-day strike), and short-term thinking (they spent too little on R&D and on new plant and equipment). In contrast, through good times and bad, Deere doggedly stuck with a plan that allowed them not only to catch their old rivals, but to leave them in the dust.*

In its seven-year development of the New Generation of Power four-cylinder tractors, Deere may have spent too much time and money, but no apologies were needed later. The product was an instant and lasting success. Behind the scenes, the changeover was done very efficiently. The Waterloo tractor factory shutdown lasted only five months, and layoffs were staggered. Relations with the United Auto Workers union stayed peaceful. The timing of the shutdown also worked well. It coincided with a year that was slow for tractor sales in general, and the introduction that followed in 1960 came as the farm economy picked up.

D-Day in Dallas was staged with the choreography of a Broadway show and the logistics of a military invasion. Hewitt staged the biggest industrial airlift the world had ever seen. Chartered aircraft delivered more than 5,000 invited dealers and industry representatives at Dallas on Monday, August 29, 1960. Another 1,500 invitees came by bus, railroad, and automobile. They came from across the United States, Canada, and 19 other countries. They included distributors, dealers, industry leaders, financiers, Deere management, and personnel. More than 100 charter buses traveled from the airport to 21 hotels and motels. Cowgirls, cheerleaders, and drill team members in miniskirts, armed with placards, guided passengers from the planes to the buses. Not a piece of luggage was lost.

Tuesday morning, August 30, the event began with a two-hour show on ice in the Dallas Memorial Auditorium. A five-sided movie screen was suspended from the ceiling. Hewitt and other speakers were televised "live" on the screens. The history of Deere & Company was shown, followed by a history of the New Generation of Power that was to be unveiled.

Hewitt opened the event with a welcoming address.

These new machines have been designed from the ground up to meet the needs of your farmer and industrial customers. Product research, design, and development have been fundamental to our progress. Six years ago at Waterloo, Iowa, we created the most modern tractor research center in the industry. Staffed with skilled engineers and equipped with complete testing facilities, this center was given the job of creating an entirely new line of tractors tailored to the exacting needs of today's and tomorrow's agriculture.

With tractors so new, so different, so completely attuned to the needs of the market, we have chosen what we believe to be an appropriate slogan: The New Generation of Power.

After these remarks, Hewitt went to the prestigious Neiman Marcus department store in downtown Dallas. At noon, the president of Neiman Marcus walked up to a 20-foot gift box and tugged away the wrapping. *Forbes* magazine described the event: "Its contents: a rakish-looking, grass-green, farm tractor. From its sides myriad diamonds twinkled the name of its maker: John Deere." A smiling Texas cowgirl was perched on the Model 3010 diesel row-crop tractor, and a diamond coronet was fixed to the exhaust pipe.

Lunch for the more than 6,000 guests was served at the Texas State Fairgrounds. After lunch, inside the Dallas Livestock Coliseum, the New Generation tractors were paraded before an enthusiastic crowd. When the parade was over, guests could see, touch, and examine the new products in the 15-acre outdoor exhibit area beside the Cotton Bowl. It was not a small parade. The show included 136 new tractors in all of their variations and 223 pieces of equipment, both agricultural and industrial, from the Waterloo and Dubuque, Iowa, factories of Deere & Company.

That evening, the guests enjoyed a memorable Texas barbecue that included 5,000 pounds of beef, open pit barbecues big enough for a whole cow, 1,800 pounds of ribs, 4,200 chickens, plus appropriate amounts of beans, salad, corn, coleslaw, biscuits, turnovers, tea, and coffee. The popular Al Hirt Band from New Orleans played the Cotton Bowl that night, followed by a fireworks show that was upstaged only by the diamond-covered tractor.

Wednesday, the guests went home to sell tractors. A gush of publicity for John Deere appeared in the news media, and a priceless new image was formed—with a glitter of diamond in the paint. The modest Midwestern farm machinery company achieved instant and international name recognition. Approximately two years of planning had gone into the $1,250,000 event. Hewitt had bet the farm on the New Generation.

The New Generation tractors were introduced a day after the Dallas event at the Minnesota State Fair. A retired engineer, Jack Hoffman, recalled:

Tish Hewitt closely inspects the John Deere name on this floor model in the Neiman Marcus department store in downtown Dallas, Texas. Each letter glittered with real diamonds. This was the day the New Generation tractors were introduced to the world after seven years of secret design, development, and testing.

That was a fantastic day. There were people ten-deep, all day long, in front of every one of them. It was unbelievable. The day before, we were sitting at a table having lunch and there were two International Harvester guys. They were discussing rumors they had heard and were having a big laugh over the fact that somebody said Deere was introducing four- and six-cylinder tractors that were going to have power brakes, etc. They just were hee-hawing, whoever thought of such a crazy thing that John Deere would do. Anyway, it was an interesting couple of days.

There was one key question to be answered: would the new product fulfill such high expectations? The measure of success is in the results. A year after the introduction of the New Generation, the factory service manager for Waterloo Works, Fred Hileman, was invited to speak to company marketing people in a tent at Waterloo's airport. He carried a small black satchel and placed it on a table beside the podium. He announced, "In this bag are one of each of all the replacement parts needed for 4010 and 3010 tractors this past year, except one. And I want to tell you about that one!" It was the differential. Full load testing of first gear in a production transmission had revealed a structural weakness in the housing of the differential gear carrier. Revised parts were already being produced.

Over the next three years, dealers sold about 45,000 of the 3010 tractors (without diamonds) and 40,000 of the 4010 tractors. The two models remained in production without major updating for eight years.

New Generation Tractors

There were four entirely new tractor models in the New Generation of Power line. The larger pair, Model 3010 and Model 4010, were built at Waterloo. The smaller tractors, the Model 1010 and Model 2010, were built at the newer factory in Dubuque.

All four had high power-to-weight ratios. Three were equipped with four-cylinder engines and power steering, and one had a six-cylinder engine. The new engines

Waterloo Works only built 170 industrial units of the 4010. This look-alike is in the Keller Tractor Collection at Brillion, Wisconsin. It has the original cab, but note the "soft" nose of the standard 4010. *Bruce Keller*

had a smoother power output across the range of engine speeds than the two-cylinder tractors. Deere advertised them as "variable speed" engines. Farmers could gear up and throttle back, reducing engine noise and increasing fuel efficiency.

The three larger models had a Syncro-Range transmission and closed-center hydraulics. The transmission provided eight forward and three reverse speeds. The closed-center hydraulic system provided up to three independent "live" hydraulic circuits serving a rear rockshaft and one or two remote cylinders.

Science and art went into the design of the operator station in a New Generation tractor. For the first time anywhere in farming, operator comfort was a high priority. Henry Dreyfuss and Associates determined the placement of the seat on the tractor from its relationship to the controls. The "comfort seat" itself was a technology breakthrough, providing orthopedically sound comfort. It could easily be adjusted to the shape and size of the driver. Suspension gave it about four inches of vertical travel to soak up road shock, with a knob for the driver to adjust the stiffness of the suspension.

This restored John Deere 4010 row-crop tractor reflects some post-production changes (such as the aftermarket cab) but was in nearly new condition and ready for sale. Restored and owned by Norman and Clifford Friesen, Eden, Manitoba, friends of the author.

The seat assembly consisted of three cushions—the upper back, lower back, and bottom. The lower back cushion fused with cushioned arm rests. The cushions were designed with the aid of Dr. Janet Travell, a posture specialist who later became White House physician for President John F. Kennedy.

The four new machines also introduced a new number identification system for John Deere tractors. Two-cylinder tractors had been identified by single-letter alphabet names when introduced in the 1920s and 1930s. These Letter Series tractors were followed by Number Series tractors in the 1950s, going from two digits (70) to three digits (720, 730).

That changed with the New Generation series. Hewitt said, in effect, we're coming out with a completely new set of models and an assortment of horsepower capabilities. Let's designate them in some logical sequence.

Four numerals were better than three. The initial size range was treated as 1000. The first model in the series would be Model 1010. It was easy to say as two words, "ten ten." When it was time to replace that in a few years, it would be upgraded to a Model 1020. If the company decided to build a midsize tractor between two series, it would take a third digit, such as 1510 or 2510.

Model 1010

In the agricultural equipment business after WWII, it was helpful for dealers to have an extended line of tractor models. The dealership might only sell one or two in a year of some models, like a small orchard tractor, but having one in the tractor row helped them to be competitive and to sell other tractors.

The New Generation 1010 series had this entry-level attraction for a row-crop tractor. The gasoline-engine model was rated at 36.13 PTO horsepower, which rose to 35.99 horsepower for models equipped with the diesel engine (a propane engine was not offered). Over five model years, 1961–1965, the 1010 tractor was manufactured in nine variations by the Dubuque factory. The total production of the Model 1010 was 44,377 units. Most were agricultural tractors, including 2,464 utility tractors and 72 orchard-type tractors. The number also included 15,803 industrial crawlers and 3,783 industrial wheel tractors.

List prices ranged from $1,866 for the crawler version to $2,292 for the row-crop utility. The diesel engine option was $500 extra. Other options included a live PTO, power steering, lights, fuel gauge, single remote cylinder control, single hydraulics with three-point hitch and swinging drawbar, or dual hydraulics with a three-point hitch and drawbar. It could have a deluxe seat that easily adjusted to the weight of the operator, arm cushions, a cigarette lighter, horizontal muffler, and power-adjusted rear wheels.

The engine for the 1010 (and for the larger 2010) was described in a 1963 souvenir book for Deere's Dubuque Works: "The very heart of each tractor is the Dubuque built four-cylinder variable-speed engine with full-load working range of 1500 to 2500 rpm—widest on the market." The engine had a vertical valve-in-head design with a four-stroke cycle. It had been preceded at Dubuque by multi-cylinder integral bore engines a year earlier for use in combines and stationary power units.

The 1010 single row-crop tractor was narrow and short by many standards. The wheelbase was only 70 inches, with a width of 53.5 inches. Overall length without a three-point hitch was 107 inches, just over nine feet. Shipping weight was listed at 3,085 pounds. The comfortable seat could be adjusted to an "offset" position, rather than centered, to permit the operator a "straight ahead" view of the row.

The row-crop utility configuration was much wider and longer. It was nearly 300 pounds heavier, 14 inches longer, and up to 32 inches wider in stance. It was 73.5 inches wide with the standard straight front axle, and long axles made it more than 83 inches wide. A long swept-back axle could also be ordered. This shortened the wheelbase for easier maneuvering. The row-crop utility 1010 also offered a PTO choice, either a single transmission-driven 540 rpm PTO or an independent dual-speed 540 and 1,000 rpm PTO.

The 1010 utility tractor had only 14 inches of ground clearance. It sat about 6 inches closer to the ground than the previous two. It was about 7 inches longer than the single row-crop but only 53 inches wide. This combination made it a handy choice for orchards or vineyards, and gave it a lower center of gravity for work on hillsides.

The 1010 was available as an agricultural crawler. The twin tracks had either four or five rollers, shoes from 10 to 14 inches wide, and a track tread of 48 inches. The crawler weighed a hefty 4,600 pounds and was marketed as a "go-anywhere, anytime" machine with great lugging capacity. It could pull a four-bottom plow (one more than the row-crop version), could work in virtually any field conditions, and was stable on rough ground, hillsides, mud, and snow. The 1010 crawler was capable of tillage and also handled dozing, loading, land clearing, subsoiling, excavating, logging, and orchard or vineyard operations.

Finally, the grove and orchard model of the 1010 had an extra-low profile and an array of shields. It was a little over four feet tall at the top of the hood. The exhaust was under the tractor. It had low-mounted headlamps, wide fenders, and side shields for the wheels.

Model 2010

The 2010 row-crop tractor offered a maximum 45 PTO horsepower in the gasoline version, using the same cylinder bore and stroke as the 1010 diesel engine. The propane 2010, with the same 145-cubic-inch displacement, used a slightly higher compression to achieve the same horsepower. The diesel version had 165 cubic inches for displacement and a 19:1 compression ratio, compared to 7.6:1 for the gasoline engine. Over five model years, 1961–1965, the 2010 was manufactured in eight versions by the Dubuque factory. Total production of the Model 2010 was 55,397 units. The row-crop and row-crop utility were the most popular by far, but there were brisk sales in the special row-crop utility diesel, in the industrial crawler and industrial wheel versions. Low-production versions were the agricultural crawler (159) and the industrial forklift (317 total).

List prices ranged from $2,563 for the row-crop utility gas to $3,675 for a hi-crop diesel. Popular options included an adjustable wide front for the row crop, independent live PTO, power steering, fenders with lights, a deluxe seat, a three-point hitch, and a rockshaft.

This is a rare 2010 agricultural crawler, fully restored, and in the Keller Tractor Collection in Brillion, Wisconsin. *Bruce Keller*

The 2010 row-crop tractor was substantially larger than the 1010. The wheelbase was 90 inches (versus 70 for the 1010) and its width was 86 inches (versus 53.5). Shipping weight was 4,600 pounds (versus the 1010's 3,085 pounds).

Marketing billed the 2010 as able to handle a three-bottom plow, four-row cultivator, all kinds of hay tools, mounted PTO, and drawn and integral tools, offering front-wheel assemblies for all crops and jobs. "It's top choice for the one-tractor farm or as a second tractor for big-scale operations." The choice of front-wheel assemblies included single wheel, dual wheels, dual wheels with Roll-O-Matic, or wide adjustable front axle. Rear wheel options were a power-adjusted set that could span 64 to 88 inches, or a "regular" option with an even wider adjustment range, 56 to 93 inches.

The 2010 row-crop utility offered an extra-low center of gravity compared to the regular row-crop 2010. Ground clearance and hood height were 4 to 5 inches lower. The wheelbase was more than 8 inches shorter and about 15 inches narrower, but it had the same shipping weight as the regular row-crop 2010. It also had different wheel packages. The choice of front axles included a straight front axle with adjustments from 50 to more than 79

inches wide, or a swept-back axle design offering extra-easy maneuvering. The rear wheel assembly could be power-adjusted from 56 to 80 inches wide, or adjustable steel disk drive wheels providing either 56 or 76 inches of tread.

Growers of vegetables, flowers, and bedded crops favored a very tall version of the 2010 known as the hi-crop. With standard 38-inch rear tires, the ground clearance was nearly a yard, 34.5 inches. The extra steel increased the shipping weight to 5,300 pounds. Tread and axle arrangements were geared toward growers who needed flexibility with an option to cultivate late in the growing season. The front and rear wheels could be set for any row width commonly used, from 54 inches to more than 89 inches. The large rear wheels and oscillating front axle eased the driving and comfort on rough ground.

The 2010 agricultural crawler had five rollers, shoes from 10 to 14 inches wide, and a track tread of 48 inches. It weighed a hefty 8,490 pounds with the diesel engine— about 3/4 ton more than the 1010 model. Both were marketed as "go-anywhere, anytime" machines with greater lugging capacity than the wheel tractor versions. Where the 1010 had a conventional four-speed transmission, the 2010 had a constant-mesh transmission with four speed ranges and, in each range, high, low, or reverse settings. Steering was also different, using a clutch-brake system with multiple-disk clutches and contracting-band brakes. The 2010 power takeoff was live, independent, and ran at 1,000 rpm rather than 540 rpm as on the smaller crawler.

Industrial applications for tractors were identified and pitched from the Dubuque factory with a 2010 wheel tractor and 2010 wheel forklift. The beefed-up industrial machines had heavier components. Factory literature described the industrial tractors as being "distinguished by modern, 'square-cut' profile and clean cut design, which give the very appearance of strength." Tractor buyers could choose from two transmissions, two clutches, and power or manual steering. The tractor had a low-profile design for maximum stability. A heavy-duty loader was available. It could accommodate either a center-mounted or five-position backhoe. Shipping weight with the diesel engine was 5,105 pounds. The forklift version came with either a 14-foot or 21-foot mast that tilted 7 degrees forward or 12 degrees to the rear. A safety canopy for the operator was standard on the tall models and optional for the other.

Model 3010

The row-crop diesel Model 3010 was an immediate and ongoing success for Deere & Company in the early 1960s, exceeded only by the larger Model 4010 diesel. These breakthrough tractors changed the face of farming and vaulted the company into the leadership position among agricultural machinery manufacturers. The four-cylinder 3010 engine was available for gasoline, diesel, or liquid propane (LP). Horsepower ratings were 55.1 for gasoline, 55.4 for LP, and 59.4 for diesel. The 3010 was available in North America for three model years, 1961 to 1963, with a total production of 45,222 units. Buyers were offered a choice of three tread versions—row-crop, row-crop utility, or standard.

More than half the 3010 tractors (23,675) were configured as diesel row-crop tractors. Another 12,525 tractors were sold with gas engines for row-crop work, and 2,442 were sold with LP engines. In the standard configuration, dealers sold 3,017 with diesel engines and fewer than 600 with gas or LP engines. Production of the five other configurations amounted to sales of 2,067 tractors. Only 76 orchard-style units were built. The crawler wasn't offered, but Deere sold 840 yellow industrial-wheel 3010 tractors. It exported 118 special row-crop utility 3010 tractors with diesel engines, and shipped another 139 to Mannheim, Germany, as the Lanz Standard.

The prices for the 3010 ranged from $4,366 for the top-of-the-line standard diesel down to $3,525 for a row-crop gas. Popular options included the wide adjustable-tread front axle ($195), rear rockshaft ($212), three-point hitch ($155), live PTO ($135), deluxe seat ($39), and air pre-cleaner ($11).

The 3010 engine used the same design as smaller engines from Dubuque, but displacement rose to 201 cubic inches with 4x4-inch bore and stroke on the gasoline and LP engines. The diesel engine had slightly larger cylinders and more displacement, almost 254 cubic inches.

Deere marketed the 3010 row crop as the "complete answer" to all four- and six-row planting and cultivating requirements, one that was able to handle large integral, power-drive, three-point equipment at maximum working speeds. The tractor had new comfort and convenience features, providing the "ultimate" in riding comfort, easy mounting, an uncluttered platform, and roomy servicing.

The new variable-speed engine, for the first time, could deliver full power at any throttle setting. The farmer could gear up and throttle back, as required, reducing noise, increasing fuel efficiency, and matching power output to the job. The new synchronized transmission, the Syncro-Range, was easy to shift through a set of ranges that provided exact speed and power for jobs from 1.5 to 19 mph. The hydraulic pump was beefed up with a large reservoir and had three live circuits available. The new Quik-Coupler enabled operators to attach or detach three-point equipment quickly and without leaving the tractor seat. A new concept in power steering made it amazingly easy to steer. New power brakes were introduced, making tight turns possible with little effort. The

This is a standard (or fixed tread) version of the 4010 tractor. About 20 percent of the 4010s sold were in this configuration. During the model's three-year production run, most of the machines sold were row-crop tractors. This machine was found in the Gordon Gilchrist Tractor Collection in Wainwright, Alberta.

John Deere row-crop tractors were quite adaptable. They could be ordered with the triangular configuration, or with the two front wheels in the older Roll-O-Matic configuration, or they could be equipped with an adjustable front axle. The adjustable axle stretched the stance from 48 to 80 inches.

The driver's-eye view on the 4010. Features to notice here include the green dash (some were later painted black), the Syncro-Range transmission (right side of the steering column), foot clutch on the left, and double brakes on the lower right. Controls for an attached loader, using the new hydraulics, are to the right of the steering wheel.

turning radius for the 5,820-pound (shipping weight) tractor was less than 104 inches.

The 3010 row-crop utility was low-built, compact, and had excellent stability. It had 18 inches of ground clearance, and a wide range of front and rear wheel tread adjustments. Operators could adjust the wheelbase from a long 92 3/4 inches to a short 81 1/2 inches. Operator comfort and convenience were high priorities. Deere literature described the 3010 row-crop utility as having a high, centered seat with an excellent view, controls that were placed for fast and easy operation, and a lighted, easy-reading instrument cluster on a slanted dash.

The 3010 standard was intended to fit the needs of big grain and rice farms rather than row-crop production. Design features included wide rear fenders, a husky oscillating front axle, and large rear tires. On standards, the rear wheels could be moved out quickly to provide greater tire-to-fender clearance for working in mud or snow. On a slightly shorter wheelbase, the model carried an extra 350 pounds and required a few inches more turning room, but it still was well-suited to extremely short turns and very stable on hillsides or rough ground. The "bull low" ground speed could be as slow as 1.26 mph, while transport speed went up to 18.54 mph.

Model 4010

Among all the New Generation tractors introduced at Deere Day, the flagship was the six-cylinder Model 4010. The model was available in three farm configurations—row-crop, standard, and hi-crop, with a choice of diesel, gasoline, or LP engines. The 4010 was marketed for three years, 1961–1963, before it was replaced by the upgraded version, the 4020. In that time, the Deere factory at Waterloo manufactured 57,573 units. Diesel row-crop units accounted for 36,736 of these, followed by 11,370 diesel standards. There were 187 hi-crop units. Deere built 170 industrial wheel versions and exported another 36 "special" standard units of the 4010.

The 4010's engine was similar to the one in the 3010. Its block was extended to make room for two additional cylinders. The 4010 had Deere's first six-cylinder wet

This 4010 row-crop tractor has one of several aftermarket cabs that were built for New Generation tractors. It has rubber mounts that deadened sound and vibration, giving a level of operating comfort that extended the work day.

sleeve, valve-in-head engine. The cylinder displacement was the same as that in the four-cylinder 3010 engine.

The 4010 made 25 horsepower more than the 3010, testing out at 84 horsepower in the Nebraska Tractor Test with the diesel engine. The 4010 gasoline and LP engines provided 80 horsepower. These choices represented a new level of row-crop tractor power, and had a retail price only slightly more costly than the 3010 tractors. Growers who didn't need that much power could stay with the modest 1010 or 2010 row-crop tractor option.

The 4010 powerhouse tractor sold for $4,116 in the no-frills row-crop version. The standard was $4,462 and the hi-crop was $5,085. The diesel engine option added $700. An LP engine cost $300 more than the base price. Popular options included an adjustable tread front axle, rear rockshaft, three-point hitch PTO, single or dual remote control valves, and a deluxe seat.

The 4010 row-crop was billed as John Deere's largest-ever row-crop tractor. The six-cylinder had the power to make full use of a five-bottom plow. It could handle a disc harrow up to 18 feet wide, or an eight-row planter at up to 7 miles per hour. Its shipping weight was 6,525 pounds. Overall length was 150 inches—12 inches longer than the 3010—but the turning radius was held to 10 feet.

The view from inside the early tractor cab was sharply restricted compared to integral cabs built 30 years later, but still provided welcome relief and protection from weather conditions.

This rear view of a 4010 diesel restoration shows some of the popular features.

The comfort seat on a vintage 4010 diesel tractor owned by Brent Gordon at Souris, Manitoba. The seat placed operators of any physical size in a natural, comfortable position. It was fully adjustable for weight and height.

The rear of a working 4010 with a swinging three-point hitch assembly and a set of hydraulic cables ready to attach to an implement.

41

"Elvis Presley had a tractor?" I heard this several times as I searched for the scoop on Elvis and his tractors. The subject came to the public's attention in early 2009 when a shining green-and-yellow 1963 John Deere 4010 tractor was unveiled at Graceland. The restoration had been a joint project of Elvis Presley Enterprises (EPE) and Deere & Company. An unlikely pairing, two of the biggest names in hardcore entertainment and hardcore machinery, but there it was.

Everybody knew Elvis was a poor country boy from the South. They knew about Graceland, the mega-tourist attraction that's like one of the Seven Wonders of the Modern World. In a lifetime, if you're lucky, you can walk around the base of the Great Pyramid, or the floor of the Parthenon, or take a boat below Niagara Falls, or do an escorted tour of Graceland.

Behind the scenes, Graceland had an old, dusty, and rusty green tractor doing a bit of yard work. If a visitor noticed it, the tractor probably seemed out of place in the manicured, candied-up 13.8-acre extravaganza of mansion, trees, and lawn in south Memphis. That John Deere 4010 is a row-crop diesel that was built at the John Deere Waterloo Tractor Works in Waterloo, Iowa, and shipped to the Planters Tractor John Deere dealership in Tunica, Mississippi, on March 12, 1963. Tunica is in the heart of the Delta Blues country, about two miles east of the Muddy Mississippi and on the fringe of a growing casino and resort area. The little town, population around 1,200, is about 45 miles southwest of Graceland. The dealership changed hands in 1985 and continues today as Parker Tractor.

Elvis Presley's 4010 Tractor
with 46A Loader

SPECIAL EDITION

1/16 DIE-CAST METAL

Partnering with Elvis Presley Enterprises, Inc. and John Deere, Ertl is proud to introduce this highly-detailed, die-cast metal, 1/16 scale replica of Elvis' John Deere 4010 tractor with 46A loader.

Purchased in 1966, Elvis enjoyed operating the tractor on his Mississippi ranch and it was later used to maintain the grounds at Graceland. Now fully-restored, the tractor is on public display in the Elvis Presley Auto Museum at Graceland.

This special-edition replica will be available for sale at authorized John Deere dealers nationwide, on ShopElvis.com and at Graceland in November 2009.

Features:
– ALL NEW die-cast front 46A loader
– Die-cast front & rear wheels
– Movable 3-point hitch
– Dash levers & PTO lever
– Rear light
– "Elvis Presley 4010 - 2009" collector insert
– Elvis Presley signature imprint

Actual Restored Tractor Shown
Replica Size is Approximately 13" x 5.5" x 5.8"
45167
SRP $79.99

Learning Curve Brands, Inc. | 1.563.875.2000 | Fax 1.563.875.5674 | www.rcertl.com

The Elvis Presley tractor is one of 57,573 units of the 4010 model that were built in a three-year run (1961–1963). There were variations during the production run, with choices for three types of fuel (gas, diesel, or propane) and three body configurations for American farms (row-crop, standard, or hi-crop). The tractor acquired by Elvis is listed by the internal production code as a Series 213, meaning it was configured as a row-crop tractor with a diesel-fueled engine. Factory records indicate that Deere built 36,736 units of the Series 213 tractors. The diesel engine produced 84 horsepower.

The tractor was equipped with a 46A John Deere loader. The 46A was a light-duty, single-cylinder loader unit, with a bucket, for New Generation 10 Series tractors. Over the years, few of the loaders have survived. This may be a unique model, in that the loader survived with the bucket and a manure scoop attached to the front of the bucket. The original manure scoop was equipped with seven tines. In the 1960s, the approximate cost for the loader was $537. The serial number provided by Graceland indicates the loader was purchased later than the tractor.

The Elvis tractor at Graceland was first purchased new by Jack A. Adams. The 4010 was included in the deal when Adams sold his little ranch to Elvis Presley in 1967.

Jimmy Gambill, maintenance director at Graceland and second cousin to Elvis, worked directly with the 4010 for most of three decades. After the restoration was completed, he shared some of his own boyhood memories of Elvis and the tractor with the restoration team.

"Elvis loved that tractor," Gambill told the team. "He rode on it. He played on it. He'd even strap a horse saddle across the hood and would ride a cousin around on it!"

Tractor Restoration

After nearly fifty years of faithful service, the dusty and rusty John Deere 4010 was assigned to a place of honor. Graceland decided, in 2008, to prepare the old tractor for display in the Elvis Presley Automobile Museum. Elvis Presley Enterprises replaced it with a new John Deere tractor and entered a cooperative effort with John Deere for restoration of the 1963 John Deere 4010. It included an agreement for future sales of diecast replicas.

Nothing was typical about the restoration. Students at Northwest Mississippi Community College, along with instructor Shane Louwerens, restored the tractor with guidance from John Deere. At the college, only Louwerens and one administrator knew they were working with the original Elvis tractor. Students only learned the secret when officials from Elvis Presley Enterprises and Deere & Company came to pick up the tractor and thank the students for their work. The tractor was placed in a newly prepared exhibit space on May 8, 2009, at the Elvis Presley Auto Museum on the grounds of Graceland.

Under the John Deere Ag Tech Program, Northwest College and Deere offer training to develop entry-level John Deere service technicians. In the second-year special project class, students are required to work hands-on with broken equipment so that they can learn how to deal with the variety of problems they will see as John Deere service technicians.

John Deere's Atlanta branch contacted Louwerens to ask if he would be interested in preserving an old tractor with the help of his students. Next, Louwerens' shop was inspected; he had no idea about the historic significance of the tractor under discussion.

"I felt like I was in a job interview at some prestigious company," he said. "They made sure our location was secure, and that we had the ability to do the project to their specifications. I was baffled when they asked me not to replace any parts unless absolutely necessary and any parts that were replaced had to be saved, bagged, and returned to them."

Then, only after Elvis Presley Enterprises decided that Northwest had the ability to complete the project, Louwerens was told the secret. Louwerens usually allows a few months to complete the special project. This time, he had 30 days to do the job. The students and instructor put 385 hours into the job. Louwerens said, "Every one of our projects gets a nickname. We called this one Stella."

The tractor arrived on February 17, 2009, shortly after lunch. It was in typical condition for a 1963 model. It required disassembly, pressure washing, and a great deal of detailed work. In all, only about ten pieces were replaced. Some dents and scratches were left on purpose to preserve the tractor's historical and sentimental value. Students even used jeweler's polish to restore (rather than replace) the original gauges and light covers.

The students were required to keep daily journals of work performed on the tractor. Louwerens took photographs to document the entire process. For collectors, he recorded serial numbers for the tractor (4010 2 T 50686) and the loader (E046A SN 10927).

John Deere and Elvis Presley Enterprises licensed the Ertl company to make a 1/16 scale diecast replica of Elvis's. John Deere 4010. The special edition replica is available at authorized John Deere dealers, at Graceland, and through toy tractor dealers. Everything on the replica is true to life, including the dents it acquired and even the unique 46A loader that is rarely found in diecast reproduction.

To this day, the original John Deere 4010 "Elvis" tractor sits permanently displayed in a place of honor, bright, shining, and polished for Graceland visitors to appreciate.

Overseas Tractors

Germany

The first two tractors introduced by Deere & Company at the old Lanz "Bulldog" factory in Mannheim, Germany, were the four-cylinder, 10-speed 28-PS 300 and the 36-PS 500. These were introduced early in 1960, before Deere Day. They were followed in 1962 by the two-cylinder, 6-speed 18-PS 100 and the four-cylinder 10-speed 50-PS 700.

The line was restyled in 1963 and broadened with two new models, the 303 and 505. The Lanz name was dropped outside of Germany. Engine side-screens were removed and small fenders replaced full fenders. The two-cylinder was upgraded to the 25-PS 200. A new four-cylinder engine from Dubuque replaced the German engines in the new 32-PS 310, 40-PS 510, and 50-PS 710.

Spain

Lanz had opened a factory near Madrid, Spain, in 1956. It continued building the Lanz 40-PS model until the Deere-designed 505 replaced it in 1963. This was followed by the 515, 717, 818, and, three years later, by an orchard/vineyard version of the 515.

Argentina

Production of the Waterloo-designed two-cylinder Model 730 tractor switched to Deere's new $3 million Rosario Works factory in Argentina as soon as it opened in 1958. It had a capacity for building 3,000 tractors a year. The factory built more than 12,000 Model 730s, in four configurations, by 1965, and continued producing them until 1970. The factory also began building an Argentine version of the Model 435 two-cylinder two-stroke diesel tractor in 1963. Using a General Motors diesel engine, these Model 445 tractors were available in four styles for regular work, row crops, orchards, and vineyards.

Mexico

A John Deere sales branch was opened in Mexico in 1957. An assembly factory in Mexico was soon building

At the Fred Haar Company shop in Freeman, South Dakota, service technician Ben Klunder is assembling a piece of harvesting equipment for a customer in the early 1960s. *Al Haar*

the two-cylinder models—the 435, 630, 730, and 830. This practice continued for successive generations of John Deere tractors through the 1960s and 1970s.

Field Equipment

The best-selling piece of John Deere tillage equipment in 1960 was the integral (three-point hitch) plow, the 810A. It was used with Waterloo two-cylinder tractors. The 415A and 416A sold nearly as many units. The plow was used with the Dubuque tractors. Integral plows had two-thirds of the market in 1960, but were on a long, slow decline. Alternative integral plows were the two-way moldboard 825 and the 475 disk plow.

The draw-type 100-series chisel plow averaged sales of 2,000 units a year in the 1960s. They were popular in wheat fields. Most had twisted double-point chisels; some had sweeps from 6 to 20 inches wide.

For irrigated corn, another primary tillage tool was the "bedder," used for preparing the seedbed. It could be front-mounted, hitch-mounted, or drawn. A grower who needed deeper soil work could use a subsoiler that would penetrate 16 to 26 inches into the ground to break hardpan.

Field cultivators and rotary cutters were in the Dallas lineup, though they would become more popular later. The CC-A field cultivator was one of several field tools made in Wisconsin at the home of the original Van Brunt

The automatic 24-T twine-tying baler was introduced in the 1960s for New Generation tractors. More than 40 years after it was built, this 24-T still sees regular work each haying season in southern Manitoba. It's owned by Philip Isaak of Rapid City, Manitoba, and is attached to a 4020 tractor.

brothers' grain drill. Horicon Works had been part of Deere & Company since 1911. Deere soon replaced the CC-A with the C-10 and C-11 integral models, as well as the C-20 and C-21 drawn models. The excellent three-point hitch on New Generation tractors from Waterloo made the integral field cultivator more popular than drawn models.

Harrowing was required on just about every commercial farm in the 1960s and 1970s. Deere & Company made good harrows. Between 1960 and 1967, Deere dealers sold more than 30,000 sections of spike-tooth harrows every year. More than 20,000 sections of a wheel-carried spring-tooth harrow were also sold annually in the same period.

Planters

The 25-B planter for cotton, corn, and other crops was a popular choice for the New Generation tractor owners. Many growers attached two of these planters to a toolbar mounted on the three-point hitch. More than 4,000 were sold each year from 1960 through 1967.

Sprayers

When the planter came off the toolbar, it could be replaced by the 25-A sprayer. This was the leading seller

in the United States from 1960 through 1967, with one sprayer sold for every two planter units. The sprayer was simple, economical, and offered a choice of six different PTO-powered pumps.

Rotary Hoes

More speed and power in the New Generation tractors gave a boost to the rotary hoe business. Speed helped the steel-tine wheels break crusted soil and flip out small weeds. Deere sold more than 30,000 one-row sections each year between 1960 and 1967. The four-row, 14-foot integral rotary hoe was the favorite.

Cultivators

New Generation tractors encouraged farmers to use wider cultivators mounted in the rear of the tractor rather than the front. With lower costs and easier mounting, farmers rapidly switched to mounting cultivators on the three-point hitch.

Mowers

The No. 10 side-mounted mower joined the line in Dallas, beside the No. 8 caster-wheel mower and the No. 9 three-point hitch mower, both introduced in 1958. A year later, the No. 11 trail-type mower was introduced. In the early 1960s, mowers were the most popular hay tool. Approximately 17,850 were sold in 1962.

Hay Conditioners

The John Deere No. 1 crimper hay conditioner had sales of 3,871 units in 1960. It used one large and one small crimping roller. The No. 2 swath fluffer, introduced in 1959, sold about 2,000 units a year as a simple, lightweight tool. A third hay tool, the No. 31 hay crusher, joined the lineup in 1961.

Square Balers

The needs of hay growers were met by square baler developments in this transition time. Deere was first to offer a square bale ejector in 1957. In 1962, Deere replaced the basic 14-T baler with the 24-T and its first bale stacker,

John Deere pull-type combines in the 1960s and 1970s were suited to large-acreage farms with low-value crops. First of these for New Generation tractors was the 96 PTO combine, introduced in 1963. Later PTO combines were the 106, 6601, 7721, and 9501. The 6601 was built between 1969 and 1984.

No. 52. Mounted on a tractor in place of a loader, the 12-foot basket could build stacks up to 24-feet high. The company also had a heavy-duty baler, the 214-T twine-tie model, and the 214-WS, which featured a single-twist knotter. Bale end size was 14x18 inches.

Choppers

Deere engineering created two flail-type machines to aid dairymen, releasing them in 1961. The six-foot No. 16A rotary chopper had more capacity, and chopped hay to shorter lengths, than the machine it replaced. Both could chop stalks, weeds, or light brush as well as standing hay.

The Year-A-Round Cab Corporation of Mankato, Minnesota, was one of several manufacturers offering cabs for New Generation tractors. This well-worn unit is mounted on a 4020 row-crop tractor owned by Philip Isaak of Rivers, Manitoba.

More than 3,000 were sold in 1961 alone. A similar rotor and knife design was used in the No. 26 flail shredder introduced in 1961. Primarily, it was used to chop cotton stalks, cornstalks, and other crop residue.

Windrowers

The self-propelled 215 grain windrower was introduced in 1961. It soon outsold the 16-foot 190 PTO-powered windrower. The 215 offered the advantage of being self-propelled and options for 10-foot or 16-foot windrows.

Pull-Type Combines

Sales were slipping for the seven-foot Model 30 PTO combine in 1959. It would be phased out in 1961 and replaced with the larger Model 42 PTO combine and Model 40 self-propelled combine. The 42 gave buyers a nine-foot grain platform for small grain or soybeans. In addition, for the first time, they could harvest corn with the two-row 205 corn head.

Self-Propelled Combines

Technology for self-propelled combines from Deere quickly improved in the 1950s and led to rising sales. Deere opened the 1960s by introducing its smallest self-propelled unit, the 40 hi-lo combine. Hi-lo combines had wider grain tanks than earlier units and concealed the top of the clean grain elevator in the tank. In 1961, Deere

John Deere became recognized as a major manufacturer of farm loaders with its 45, introduced in 1955. The 48 loader came out in 1969. It could fit any tractor from a 2010 through a 4020. This 146, along with the 148 and 158, was introduced in the first half of the 1970s. It remains an important tool on many farms.

introduced the 105-horsepower 105 combine. It featured 14- to 22-foot cutting platforms, five straw walkers, and a 75-bushel grain tank.

Cotton Pickers

Cotton-picking machines changed the way cotton was harvested in the South in the 1950s. By 1960, the Deere line included one-row and two-row tractor-mounted cotton pickers. The company built 2,461 tractor-mounted cotton pickers in 1960, including a new model for the New Generation of Power. The one-row 22 picker mounted on the 2010, 3010, or 4010 tractor, and the tractor was driven in reverse. Deere sold more than 1,000 units annually from 1960 to 1962. The all-time best-seller, however, was the Model 99 two-row self-propelled cotton picker. Annual sales exceeded 1,000 from 1960 to 1962.

European Combines

The old Lanz factory in Zweibrucken was visited in 1961 by a design team from the Harvester Works in East Moline. The team was charged with development of a new John Deere combine suitable for Europe. Many models and sizes were already being built at the factory, but major changes were required. The design team discovered, among other things, that two competitors had modified the John Deere 55 combine so that it could handle the heavier straw conditions in Europe. Discoveries by the team led to the introduction of the 30 Series combines for Europe in 1965. These included the 330, 430, 530, and 630. This began a long line of German-built John Deere combines through the 1980s.

Other Farm Equipment

A long line of other equipment for handling farm and livestock material was on show that day in Dallas. The 200 bale elevator was a one-man hay handling system, able to move small bales as a conveyor or an elevator. It became a best-seller, with sales exceeding 2,000 units a year in 1961 through 1965. The same farms were likely to buy either a new 43 power sheller or 10-A hammer mill in 1960. John Deere rotary cutters also were popular. Sales of the 127 model exceeded 5,000 in 1960 and 1961. The six-foot 307 utility cutter was introduced in 1961, followed in 1962 by the seven-foot 407 Gyramor cutter. The Gyramor established Deere in the heavy-duty market with its ability to clear four-inch-diameter brush.

Deere dealers also had a selection of farm loaders, manure spreaders, rear blades, wagons, post-hole diggers, implement carriers, and even skid steer loaders in the yard or ready to order for customers as they scanned the New Generation tractors. By 1963, the company also offered a John Deere 110 lawn and garden tractor.

Restored to pristine conditions, the John Deere 3020 diesel tractor is one of the most popular of all time. It was built for model years 1964 through 1970. Gas and propane were options for the diesel engine. This is a standard tread edition, casually parked on a gravel lane one sunny spring day in southern Ontario. *Tony Gerber*

THE CLASSIC ERA: 1964–1971

The Class of '64

A new generation, the sixth since John Deere founded his company, began in the 1960s. Baby Boomers Adrienne and Anna Hewitt and their younger brother, Alexander (Sandy) Hewitt, lived on the Hewitt family estate near Moline. The three would go on to outstanding personal careers, but they would not be on the board of directors or controlling the path of Deere & Company.

Outside the Hewitt estate, society was reinventing itself. Technology was transforming the world into a global village connected by radio, television, film, and the first communications satellites. Public concerns had international dimensions, whether they were the civil rights movement, Vietnam, music, religion, famine, birth control, or drugs. The international scope of these issues had impact even in Moline and the Quad Cities.

Near the small community of Moline, Tish Hewitt was fulfilling her own dream. She had loved horses since she was a child. Before Bill was elected president, she had transformed 375 acres on a bluff overlooking the Rock River into Friendship Farms, a world-class facility for breeding Arabian horses.

Her daughter, Anna Hewitt, described the estate for a *Sports Illustrated* article in August 2001: "The farm was where we spent all our leisure time. Mother didn't push

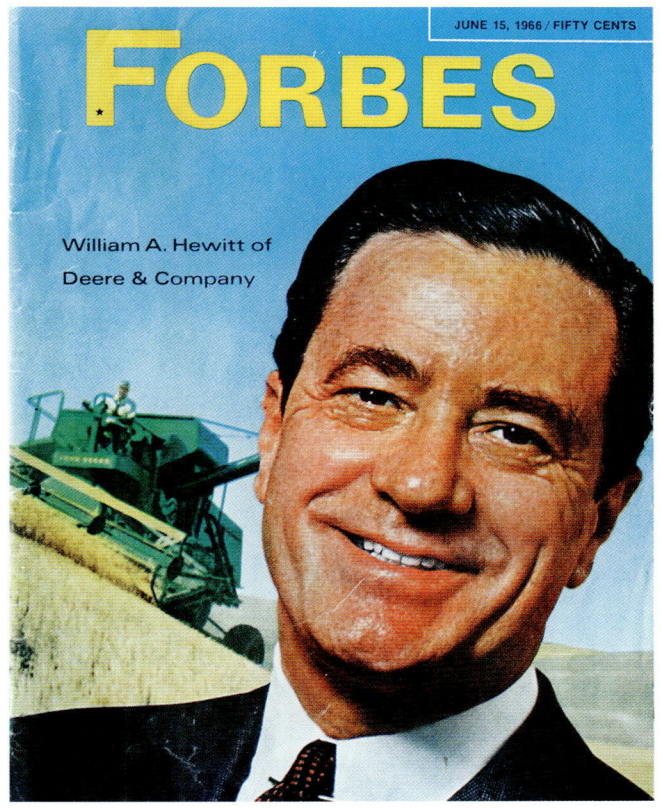

JUNE 15, 1966 / FIFTY CENTS

FORBES

William A. Hewitt of Deere & Company

riding. We would have to want to go, and we did. We would trail ride. It was a beautiful spot—the ravines, the woods, the pastures."

While Deere was bringing John Deere products to the world, Tish Hewitt was bringing the world to Moline. She

Forbes magazine published a cover story describing the success of Deere & Company on June 15, 1966. Wall Street had considered farm equipment "a cyclical industry" with a low earnings ratio for investors. Forbes challenged that on March 1 and June 1, and capped it with the report on how Deere was cashing in on a great boom in demand worldwide for farm machinery.

was described as "a force" in the Quad Cities in this era. She marched with the area's African American leaders after the assassination of Martin Luther King Jr. in 1968, and opened her home for meetings of civil rights leaders.

Under her care, Friendship Farms became a place of serene beauty. After Tish Hewitt passed away in 1992, her children decided to share the beauty of Friendship Farms in a public way. Bordered by the Rock River in Silvis, Illinois, about two miles east of John Deere World Headquarters, the 386-acre TPC Deere Run club opened in 2000 and today is the home of the John Deere Classic on the PGA tour. The course has garnered golf industry acclaim as one of the most scenic and challenging designs in the national network.

Global Outreach

In April 1964, Bill Hewitt asked his directors to change the titles of two positions. He asked to be given the title of chairman, and that the executive vice president (his closest colleague in the group), Elwood Curtis, be appointed president. Outside the company, Hewitt explained, business people felt the term "president" carried more weight than "executive vice president." As chairman, he would remain the chief executive officer rather than chairman of the board. Hewitt's strengths in general management, marketing, and public and community relations were counterbalanced by Curtis' strengths in analysis and financial acumen.

After launching the New Generation of Power, Bill Hewitt did some behind-the-scenes reshaping of the company's organization so that it could function effectively as a single company on a worldwide basis. The domestic arm of the company retained its structure: one group for manufacturing and one for marketing. Senior managers were sent to Europe to direct Deere affairs firsthand while gaining international experience. A new Canadian company, John Deere Intercontinental, was formed as a rallying point for operations outside of the United States. Two divisions were set up—one for Latin America, the Pacific, and the Far East and the other for Europe, Africa, and the Middle East. Structural changes

An executive chat between three leading figures of Deere & Company in the 1960s and 1970s. New chairman Robert Hanson (center) is shown between Bill Hewitt (left) and president Elwood Curtis (right). Hanson rose through the company ranks to become Hewitt's successor in 1982. *Anna Wolfe collection*

culminated in early 1963. Adopting most of a new report by Booz Allen Hamilton, Hewitt led a radical departure from company structure overseas.

Wayne Broehl Jr. writes,

Deere's international operations argued for a different structure. These factors were the wide variety of ownership situations, the special complexities of international business itself, the development of the 'worldwide' tractor concept, and the particular needs for coordination between France and Germany. Their conclusion: the company should be organized abroad in a geographical mode. Two senior vice presidents would be assigned responsibility for managing the two major segments of the business—the North American operations and the international operations, respectively. In turn, the international operation would be divided into two groupings each to be headed by a vice president. One would embrace all of the European operations; the other everything outside of Europe (i.e., Argentina, Australia, Mexico, and South Africa).

William Alexander Hewitt in front of the new Deere Administrative Center building. Under his leadership, Deere gained worldwide recognition for its agricultural equipment. He chose to keep the headquarters connected to its roots, in its Midwestern home at Moline, Illinois. *Anna Wolfe collection*

That basic pattern still stands today.

With strong products, strong leadership, and an effective organization, Deere & Company reaped a tidy net profit during the first 10 years of Hewitt's leadership, 1955–1964. The company led the farm machinery industry in management and profitability. It used the cash flow to re-invest in product development, capital improvements, and, ultimately, customer support.

Deere annual sales rose from $339.6 million in 1955 to $816 million in 1964. Annual profits on sales were never less than $20 million, and rose to $59.4 million in 1964, despite a downturn in the farming economy

in the early 1960s. This solid footing supported the $40 million cost to deliver the New Generation of Power in 1960, which was followed by substantial new outlays in each year through 1964. When other companies let R&D slide in this era, Deere kept improving products and production efficiency. From 1960 forward, Deere invested more than 4 percent of its sales into product research and development.

Deere & Company achieved Hewitt's vision of being number one in the industry in 1963, when it surpassed International Harvester as the world's largest producer and seller of farm and industrial tractors and equipment. Deere & Company kept that position over the next two decades.

Deere sales of farm and light industrial equipment hit $762 million in 1963. Competitor net sales that year were: International Harvester, $665 million; Massey-Ferguson, $636 million; Allis-Chalmers, $544 million. Deere's share of wheel tractors in the United States increased from 23 percent in 1959 to 34 percent in the same year. Deere also moved ahead of its farm machinery competitors in net income per dollar of sales and per dollar of total assets.

Completion of the Deere Administrative Center in June 1964 was significant to their improved market position. The center was a world-class architectural and aesthetic success that brought the finest modern architecture in virtually any city on earth to a modest Midwestern town. The complex presented a new face to the world, that of a progressive corporation representing the best in quality, style, and effectiveness. Judging by this impressive facility, this corporation's primary product could have been anything; it happened to be tractors.

Worldwide Design

Weeks after the stunning impact of D-Day in Dallas, Bill Hewitt laid out a new vision for Deere. The goal he outlined was to create a tractor for the world. The initiative arose at a boardroom discussion in September 1960 with 34 top executives assembled. This was the vision that Hewitt presented, according to Broehl:

The first John Deere Worldwide utility tractors were the little 1020 and 2020, manufactured in Dubuque, Iowa. The three-cylinder 1020 with a swept-back front axle could turn a tighter corner. It weighed 5,955 pounds with ballast and could pull a 4,200-pound load or lift 2,000 pounds. The last list price was $4,485, with model years from 1965 to 1973. Photographed in the Gordon Gilchrist Collection in Wainwright, Alberta.

Deere & Company leads competition in the United States; however, competition is ahead of us in the foreign markets . . . Investments for foreign manufacture will be rewarded if properly planned, and we can contribute to the world welfare by making equipment available . . . . We should evaluate this market, the facilities required, and production capabilities to assess the potential for a worldwide tractor, bearing in mind that farming customs and conditions differ in various countries, and it would be uneconomical and impractical to attempt to design a tractor for each country. Therefore, the final design must represent a composite of world requirements.

Out of the early discussion came plans for new prototypes that would be the heart of the Worldwide tractor program. As Broehl recalls, the mandate to engineering was that "all Deere factories everywhere in the world were to design their products with the maximum possible standardization." This would allow interchangeability

of parts and maximum use of design talent, raise quality standards worldwide, and result in tractors that looked like they belonged to the Deere family.

Engineers developed two three-cylinder diesel prototypes for the Worldwide program. The X-21 was rated for 37 horsepower at the PTO and the X-22 for 55 horsepower. After three years of development, North America's versions of the first Worldwide tractors came out of the Dubuque factory in 1965 as the Model 1020 and Model 2020. A third, smaller version was developed for the European introduction in 1965. The three tractors carried a three-digit model number in Europe (310, 510, 710) and were styled as members of the New Generation of Power family.

Models 310, 510, and 710

The X-25, the smallest of the three prototypes, became the 32-horsepower 310 tractor. The other Worldwide tractors Deere introduced were the 40-horsepower 510 and the 50-horsepower 710. These tractors gave

The 2020 was sized to be a utility tractor. The components were adapted for production in factories in Europe and Mexico, as well as America. It had an 85-inch wheelbase, 12 inches longer than the 1020, and weighed up to 7,435 pounds. Base price in 1971 was $5,800. Options included a four-cylinder gas or diesel engine. Photographed in the Gordon Gilchrist Collection in Wainwright, Alberta.

Deere a presence in two countries in Western Europe, and marked an achievement in international cooperation. The factory in Saran, France, made the diesel engines; transmissions and several other components were manufactured in Mannheim, Germany; a significant number of other components were manufactured in the United States.

While the Worldwide design idea was being integrated into production, other new John Deere tractors were introduced or updated in Europe. In 1963, in France, two tractors were released for the French market from the new factory at Saran. The 37-horsepower 303 and the 44-horsepower 505 were built by a Deere-led consortium of companies. They were German-styled with shell fenders, but bore the name "John Deere" rather than "John Deere-Lanz."

In the same year, in Spain, the first Deere-designed tractor was introduced at Spain's first John Deere dealer convention. The 505 was built at the Getafe Works near Madrid and carried some recognizable New Generation

styling features. It was fitted with a 44-horsepower Perkins diesel engine. Successors were introduced in 1966 as the 515, the 515V vineyard version, and the 717. The 515s were replaced in 1967 with a larger 60-horsepower 818. These tractors were powered by a Dubuque-designed 20 Series engine built in the new Deere factory at Saran, France.

U.S. Worldwide 20 Series

The Worldwide tractor concept began to bear fruit in 1965. When engineers completed the three-year design project, the factories in Dubuque, Iowa, and Mannheim, Germany, could use the same set of tooling. At first glance, American customers might have thought the new 1020 and 2020 utility tractors introduced in 1965 were updates to the 1010 and 2010. Instead, they were the first of the John Deere Worldwide tractors.

The Worldwide models from Dubuque shared concept features such as engine, hydraulics, transmission, and operation with the Waterloo-built 3020 and 4020, but they were better adapted for international production and marketing than the larger New Generation tractors. Thanks to the design initiative, many parts were interchangeable; this reduced the number of distinct parts required by about 35 percent. The manufacturing cost for the new models, compared to the 1010 and 2010, was 25 percent less but still met requirements for new features. Size-wise, they were suited to Europe's smaller farms and gave Deere a competitive presence against established European tractor makers.

Dubuque supplied two engines for its three utility tractors and dropped the low-production liquid propane engine. Both the 1020 and 2020 had a new two-speed foot clutch. By depressing it halfway, the operator could stop any forward motion while the PTO continued spinning, allowing an implement to continue processing heavy crop material.

Model 1020

In the 1020, a slightly more powerful three-cylinder engine for diesel or gasoline replaced the four-cylinder engine used in the 1010. The three-cylinder had 33 horsepower at the drawbar and 38 on the PTO.

In other features, the 1020 was the first utility tractor with closed-center hydraulics and a three-point hitch with lower link sensing. The eight-speed transmission had four gears in high range, low range, and reverse. The 1020 was offered in five configurations: low utility, row-crop utility, high utility, grove (with shielded fenders and a low profile), and hi-crop. All models had wide front ends.

Model 2020

The 2020's engine was a four-cylinder rated for 46 horsepower at the drawbar and 54 horsepower at the PTO. That engine was used in both the 2020 and the 2510, the last of the New Generation tractors.

The 2020 was also available with a foot-engaged differential lock, which provides improved traction in slippery conditions. The 2020 was offered in four configurations: low utility, row-crop utility, high utility, and grove or orchard tractor.

New Generation Latecomers

A tide of bright green-and-yellow New Generation tractors swept into hundreds of thousands of farms in North America in the 1960s. Twelve years passed before Deere introduced the Generation II tractor in 1973. In the intervening years, Deere supplied two more New Generation tractors and a series of upgrades to the New Generation theme.

Model 5010

The 5010 standard from Waterloo Works was introduced in 1962, during Deere's 125th anniversary. The model was advertised as the "most powerful standard tractor on the market" and was built for large-acreage grain and rice farms. The 5010 was the first two-wheel-drive tractor with 100 drawbar horsepower. Output at the PTO was 117 horsepower. The model was powered by a John Deere six-cylinder diesel engine. It also came with a Syncro-Range transmission and was the first John Deere tractor with 1,000-rpm-only PTO.

The 5010, with a new category 3 hitch, was ready to handle the heaviest equipment available. This category included implements like a 40-foot grain drill, seven-bottom semi-integral plow, 32-foot disc tiller, 34-foot field cultivator, 32-foot mulch treader, 27-foot disc harrow, 27-foot tool carrier, and similar field tools.

The 5010 also had hydraulic power brakes and power steering, a dust shield, and wide rear fenders. A larger, new tire size was an option, and owners could set it up with dual wheels. Factory-installed cabs, built by another company, were optional. Air conditioning wasn't available yet, but operators could order it with a cab heater.

More than 5,400 units were built in model years 1962 through 1965. Most settled into the big wheat farms of the Great Plains and prairies, places where farm managers needed to work big acres in a hurry, especially during a short seeding window.

The 5010 was not available as a row-crop tractor, but an industrial version was released. Deere built an additional 2,000 of these popular additions in just three years, painting them industrial yellow.

The big daddy of the John Deere New Generation tractors was the 5010 diesel, photographed at the Manitoba Threshermen's Reunion held in July 2009. This is a 1963 model owned by Glen Sawatzky of Austin, Manitoba.

The 5010-I for industrial use was made available in late 1962, with an offset operator station. The offset design enabled the attachment of fifth-wheel units like an elevating scraper. It had 60 percent more power than the 840, which it replaced. Approximately 2,000 were built. *Bruce Keller*

The 2510 tractor built from 1965–1968 was a crossover tractor with a Dubuque body and a Waterloo engine suited to the needs of thousands of farms. It had a snappy response to the throttle and good handling that set it apart. Row-crop and hi-crop versions could be outfitted with many options, including a rear quick-coupler. It was introduced in August 1965. For the 1967 model year, a Roll-Gard option and factory cab, with or without heater, were added to the options list. *Tony Gerber*

Model 2510

Customers spotted a third newcomer in the 1965 model line beside the Worldwide models. The 2510 was the smallest and last of the New Generation tricycle-type row-crop tractors; it presented a blend of technology from two factories.

The 2510 was a Waterloo-built 3020 tricycle tractor that was powered with a Dubuque-built 2020 engine. It met the need for a smaller row-crop tractor than the 65-horsepower 3020. With 54 horsepower, it still offered features of bigger Waterloo tractors such as power shift or Syncro-Range.

Several front-end options made it highly adaptable. These included the dual-wheel tricycle, regular or heavy-duty Roll-O-Matic, and a regular-adjustable or a wide-adjustable front axle. There were four models: row-crop, row-crop utility, utility, and hi-crop.

New power steering was promoted on the 2510 tractors. The single lever enabled operators to shift on the go through all forward and reverse selections without clutching. Power steering, power brakes, a power-on-demand hydraulic system plus rack-and-pinion rear tread adjustment were some of the other attractive features.

New Generation Updates

In the fall of 1963, the 3010 and 4010 were replaced by the updated 3020 and 4020. The public reception to this pair of tractors was spectacular, creating an instant headache for competitors.

Waterloo's engineers had refined these tractors in several ways. The engines had improvements in pistons, rings, cylinder blocks, and liners. An alternator replaced the generator. They used a dry type of air cleaner. The upgraded engines produced more horsepower.

The big news, however, was the optional power shift transmission. Most tractor-makers had introduced some form of shift-on-the-go transmission ahead of Deere. These allowed operators to change transmission ratios for load and field conditions. They required operating two levers, or three, and a foot pedal.

Deere's new power shift transmission in the 20 Series allowed the driver to shift with a single lever on the go, under load, without clutching, into a choice of eight forward gears and four reverse gears. The former clutch pedal was put to a new use: it became an "inching" pedal for use in hooking up implements and other close-in work. To prevent trouble starting the power shift tractor in cold weather without a clutch, Deere engineers equipped it with a lever to disconnect the engine and transmission.

The Syncro-Range transmission was still available. It was dependable, cost less, and was beefed up to handle the new model's increased power.

Model 4020

The 4020 sold more units than any other single model of farm tractor ever built by Deere & Company. Between 1963 and 1971, approximately 177,000 units of the 4020 were built and sold. The 4020 in 1968 accounted for 48 percent of all Deere tractor sales. It became known as the "Classic," a name which has since been applied to both the 4020 and 3020.

The 4020 diesel 6-cylinder engine was upgraded. Displacement was increased in each cylinder by 6 cubic inches, to a total of 404 cubic inches for the engine. In the Nebraska Test, the diesel engine produced 91.2 horsepower at the PTO, about 7 horsepower more than the engine in the 4010. The 4020 came in three body types for farming: row-crop, standard, and hi-crop. There were small variations for the export units and an industrial version, identified as the JD600.

The 4020 could be ordered with an all-steel cab built by Crenlo to provide protection from the weather. It could also be equipped with a heater and windshield wiper. The base price for a new 4020 row-crop diesel in the fall of 1963 was $5,714. Equipped with the power shift transmission, two-speed PTO, dual hydraulics, and a universal three-point hitch, it could be purchased for less than $7,000. The price was similar to or less than competitors were asking for less attractive tractors, and it came with Deere dealer support and service.

The John Deere 4020 was just a little bigger than the 3020 and ranked in some quarters as the greatest Baby Boomer–era tractor. It was big, powerful, comfortable, easy to operate, and highly reliable. Parts are still available, many are still working, and some have been restored. *Tony Gerber*

Model 3020

Deere also had spectacular sales with the 3020, introduced at the same time as the 4020. Between 1963 and 1971, Deere manufactured and sold more than 86,000 units of the 65-horsepower 3020 tractor. The 3020 engine featured a five-horsepower increase over its predecessor. The 3020 was offered in five models: row-crop, row-crop utility, standard, hi-crop, and orchard or grove tractor.

Updates with new features were provided for America's hottest mainstream tractors, the 3020 and the 4020, in 1966 and 1968.

Roll-Gard

Tractor overturns were a problem for the industry before Hewitt came on the scene. By the mid-1950s, the industry had suffered an estimated 500 to 700 tractor rollover fatalities. Both International Harvester and Caterpillar experimented with rollover protection, without much success. John Deere began experimenting with rollover protection structures (ROPS) around 1960. In 1966, after investing more than a million dollars and several years of research, the first patent was issued for a John Deere ROPS option known as Roll-Gard.

Hydraulic control levers were mounted on the left side of the upper dash area on this 1967 original condition 4020 tractor, still operated by Leonard Lepp near Rivers, Manitoba.

The 4020 has an operating dash lamp, five windows for the instrument cluster, and power shift controls.

The driver's view on this 1969 restored 4020 near Rivers, Manitoba, owned by Ted Krahn, reveals an updated design for the dash. Power shift transmission control is on the right. The PTO lever is on the left side of the dash. A keyed ignition switch starts and stops the engine.

The rear of the 4020 is equipped with a drawbar, PTO, adjustable wheel spacing, three-point hitch, three sets of remotes for hydraulics, backup lamps, comfort seat, hydraulic controls on the operator's right, and struts supporting a canopy.

This is a later, fully restored 3020 diesel row-crop with a canopy and full set of wheel weights. Restored and owned by Tony Gerber, Millbank, Ontario. *Tony Gerber*

Roll-Gard became one of the first updates offered to the 3020 and 4020 tractors from Waterloo in 1966, but it was an option that wasn't widely adopted. It became more acceptable, along with ergonomically designed cabs, in the next decade. Eventually, the patents on Roll-Gard were made available to all tractor manufacturers so that all tractor drivers could enjoy this protection.

As recently as 2001, an occupational injury survey identified the top five tractors without ROPS in the United States. It put the still-surviving John Deere 4020 and 3020 at the top of the list, estimating farms still had 105,000 John Deere 4020 tractors and 55,000 3020 tractors still active and without ROPS. The other three were Ford and Farmall tractors from the same general era.

Power Front-Wheel Drive

Deere introduced a second major feature in 1966 on the 3020 and 4020 tractors. Called "Power Front-Wheel Assist" (PFWA), it retained the flexibility and efficiency of these two-wheel-drive tractors, but assisted them when traction was poor. Using hydraulic power, PFWA gave a 20 percent boost in traction and was the forerunner of today's popular mechanical front-wheel drive.

Gerald Mortensen was working on front-wheel assist in the early 1960s. In personal correspondence, the engineer recalled testing a supplier's promising proposal for a hydrostatic front drive in about 1962. John Deere's Product Engineering Center (PEC) developed it into a unique hardware upgrade for the 20 Series tractors. The upgrade to PFWA, using the hydraulic system, also provided an

This 5020 diesel, also known as a Wheatland tractor, was a popular tractor with 121 horsepower on the drawbar. This one is equipped with an aftermarket cab and dual 24.5x32 tires. Deere's Wheatland tractor exceeded the capacity of 18x26 tires and led to the development of a new tire size that could handle the weight and drawbar power. Located at Portage la Prairie, Manitoba.

adjustable, offset front axle. This allowed the 20 Series tractor, with PFWA, medium-size front wheels, and attachments, to have a short turning radius and adequate crop clearance. At first, Mortensen recalled, some customers were disappointed that PFWA didn't provide a full mechanical front-wheel drive; however, most were delighted with the improved performance as soon as they encountered wet field conditions.

Model 5020

Customers found another updated model at American dealers in 1965, a 20 Series replacement for the 5010. From Waterloo, it was the industry's most powerful two-wheel-drive tractor in 1965.

The 141-horsepower 5020 standard was the first two-wheel-drive tractor with more than 130 horsepower. It was fitted with a 531-cubic-inch, six-cylinder engine, and weighed up to 19,630 pounds. Features and options were

similar to the 5010, including a Syncro-Range transmission, deluxe seat, power brakes, and power steering.

Between 1965 and 1972, Deere built approximately 13,000 units of the 5020, including a row-crop version that was introduced in 1967. The last base price for the 5020 was about $14,500.

Worldwide Updates: 1965-1966

Overseas in the mid-1960s, updated Worldwide tractors rolled off Deere's production lines. These new versions were the result of collaboration between American, French, and German companies. The 1020 and 2020 used American eight-speed transmissions. Their German counterparts, the 310, 510, and 710 tractors, had Lanz 10-speed transmissions. The 310 produced 36 horsepower with a three-cylinder engine; the 510 offered 45 horsepower with its three-cylinder engine; and the 710 had a four-cylinder engine and produced 56 horsepower.

Waterloo-based transmission engineer Bob Haight recalled field research conducted in the 1960s in Texas. The engineering department would hold winter test programs in January and February in southern Texas. After four to six weeks, Deere officers joined in for a few days. Bill and Tish Hewitt usually led the delegation and stayed two or three days.

[Hewitt] was a nice guy, and he was always very interested in talking to the mechanics and the people that did the evaluation work. Bill always headed right for

wherever the mechanics and engineers were standing. He wanted to know, "What's your opinion of this machine," and "What do we need to fix." He would always come over to where we were collected and he would want to know, confidentially, what's good and what's bad.

[Tish] always came down with Bill. They'd come through and operate and get "the feel" for the tractors. The group of engineers that was there always commented about it afterwards, that she was a better tractor driver than he was.

Deere factories in Argentina and Spain also came out with new models in 1965. In Argentina, Deere executives were in a long struggle with Argentina's changing operating requirements. By 1965, the Rosario factory had finished a substantial plant addition that enabled Deere to increase the content of local components to 90 percent. The factory introduced four models in 1965, from a 43-horsepower 1420 to the 2420, 3420, and finally the 102-horsepower 4420. They were manufactured through 1971.

The factory in Spain had three new Worldwide tractors available for 1965. They were the 45-horsepower 515, the 56-horsepower 717, and the 818 with 60 horsepower.

Further updates to the Worldwide series came in 1968. These included the 820, 920, and 1120 from Mannheim.

Mid-1960s Ventures

Deere had ventured into the consumer market in 1963 when it decided to build and sell lawn and garden tractors. It also began building attachments and snow blowers.

The company built a link into Asia in 1963 through a licensing agreement with Hitachi. The Japanese company would build Deere farm and industrial tractors and equipment for Japan, South Korea, and Okinawa. This

was an agreement forced on Hitachi by pressure from the Japanese government. Hitachi was more interested in other products, and the venture ended in 1970 with little to show.

An earlier venture ended in 1965, when Deere sold its only venture into producing crop inputs. In 1951, Charles Wiman had initiated the John Deere Chemical Company in Pryor, Oklahoma, to produce fertilizer. When a purchase offer came from a Texas company in 1965, the Deere & Company board of directors accepted the offer and signed off as a supplier of crop inputs.

Company sales surpassed $1 billion for the first time in 1966, and earnings reached a high of $78.7 million. Farm equipment sales set a new record for a fourth straight year, and industrial equipment sales notched their largest ever year-to-year increase. Deere & Company's lawn and garden equipment sales rose 76 percent, and worldwide employment hit a new high.

The Clean-Cut, Contemporary Look

John Deere released a new trademark in 1968 as it prepared to release a wave of tractor updates. Describing the new trademark, a company memo said, "The new trademark is in keeping with the progress being made throughout all divisions of the Company . . . it provides

Canada's dairies liked the Mannheim-built 1120 diesel tractor, made between 1966 and 1971 specifically for the Canadian market. It was a handy main tractor for dairies, able to handle a four-bottom plow, and could serve as a second tractor on grain farms. This one, equipped with a 145 farm loader, was visiting Enns Brothers in Winnipeg.

for better reproduction and greater readability under a wider range of usage."

The design was modernized and streamlined to show a straight-side silhouette of a deer with just two legs instead of four, and one four-point rack of antlers. The "John Deere" logotype was changed with a hand-modified version based on the Helvetica font. The width of the ellipse border was narrowed and the size ratio of the deer inside it increased.

The new trademark appeared on the updated 20 Series New Generation tractors released in 1968 along with several new tractor models. The new 1520 came out of Dubuque; the new 2520, 4000, and 4520 (John Deere's first turbocharged tractor) came out of Waterloo. In addition, a pair of four-wheel-drive tractors, the WA-14 and the WA-17, came in from the West Coast.

New Generation Classics, the 3020, 4020, and 5020, had updated and refined engines in 1968, with improved power output. The PTO power was increased from 64 to 70 horsepower on the 3020 and from 91 to 94 horsepower on the 4020. The 5020 was boosted to an industry-leading 141 horsepower. Oval mufflers made these later units quite distinctive.

Canada's customers had responded well to earlier four-cylinder, Mannheim-built 710 tractors with the Worldwide configuration. In 1968, Mannheim Works became Canada's supplier of John Deere utility tractors. Models selected initially for the Canadian market were the 36-horsepower 920 and the 45-horsepower 1120. The 2120 was released a year later. It had a four-cylinder engine and longer wheelbase, and went on to become a popular loader tractor on dairy farms in central and eastern Canada.

Model 1520

As the first Baby Boomers were graduating from university classes in 1968, going off to war and anti-war protests, Deere responded to the needs of many Midwestern farmers for a midsize to small utility tractor. Its three-cylinder, eight-speed, 46-horsepower 1520 could handle a variety of farm chores, from plowing to haying. In Nebraska Tests in 1968, the 1520 substantially out-performed the Worldwide 1020 tested just two years earlier. It outweighed the earlier tractor by 1,000 pounds and had about 1,000 pounds more pulling power on the drawbar. It was made in Dubuque for the U.S. market.

For an even smaller tractor, U.S. dealers could stock the new Mannheim-built three-cylinder 31-horsepower 820 tractor.

Only a few hundred hi-crop versions of the 2510 and 2520 were produced. Most of those were for sugar cane growers who needed the high clearance and power for cultivating. In the 2510 series, only 7 hi-crop tractors burned gasoline and only 132 fired on diesel. This set of three 2520 tractors, two diesel and one gas, is unique and found in the Keller Tractor Collection. *Bruce Keller*

Model 2520

The 61-horsepower 2520 row-crop was an upgrade to the factory crossover 2510. It had the Waterloo tricycle chassis of the upgraded 3020 and the four-cylinder engine of the upgraded 2020. Refinements in design improved power output by about 10 percent.

It was offered as a row-crop or hi-crop tractor, with gas or diesel engines. It was also offered with the Roll-Gard structure.

Model 4000

A second crossover tractor, the 97-horsepower 4000, was introduced in 1968. It shared the same six-cylinder engine as the newer 4020 and the same rear end as the 3020. It could handle an 8,000-pound load—more than the 3020, but less than the 4020. The 4000 had an economy version of the Syncro-Range transmission, which meant it had to be stopped when shifting into reverse. The tractor could also be ordered without the frills of sheet metal around the seat to cover the hydraulics, or battery covers.

The 4000 was meant for farmers who wanted the power of the bigger 4020 at a lower cost. As long as the machines weren't overloaded, they worked well. They were about a thousand pounds lighter than the 4020.

In the 2520 tractors, this is the key feature of a rare 2520 diesel owned by Ted Krahn of Rivers, Manitoba. It is equipped with power shift rather than the standard Syncro-Range transmission.

This 2520 row-crop diesel has a rare power shift transmission, a dual headlamp, and the small injection pump typical of the 1969 and 1970 models. Later units came with a single, dual-beam headlamp in each fender, a dimmer switch for the operator, and a larger flasher lamp on the fender. Gauges were also updated in later models.

The 2510 was created by installing a Dubuque engine into the Waterloo body. The 4000 was the result of another crossover combination. The 4000 had the six-cylinder 4020 engine and the 3020 differential. The no-frills 4000 was popular. This one is owned by Owen Brooks of New Lowell, Ontario. *Owen Brooks*

There were interesting variations of the 4000. Some were ordered with a Roll-Gard structure. The power shift transmission was an option for model years 1971 and 1972. A low-profile model was built for California's grove and orchard operators, who needed both high horsepower and a low-profile tractor with a short turning radius. There were only 46 of these built, and of that number 21 were equipped with the improved eight-speed Syncro-Range transmission.

Model 4520

Deere & Company capped its 1968 introductions with its first factory-installed turbocharger. The 4520 was powered by a turbocharged version of the six-cylinder 404-cubic-inch diesel that had proven itself in the 4020 tractors.

The 4520 was offered for two model years, 1969 and 1970, before being replaced by the 4620 for 1971–72. Both the 4520 and 4620 could be purchased with a front-wheel-drive option and with either a Syncro-Range transmission or a power shift transmission. Retail base price for the 4520 was $11,600.

Turbocharging increased the power output of the six-cylinder engine by 25 percent. The turbo's exhaust-driven fan forced air into the engine's combustion chambers, allowing more fuel to be burned and more horsepower to be generated. The 4520 produced 122 horsepower on a 1969 Nebraska Test and was rated to pull a 13,000-pound load.

The Waterloo-built 4520 was equipped with an entirely new, beefed-up frame that could handle the increased stress of turbo power. To dissipate the greater heat output, the tractor featured a larger radiator and fan, more capacity in the air cleaner, a redesigned engine block (to allow more coolant flow), and a larger-capacity lubricating system.

Big Power Competition

In the 1950s and 1960s, the tractor industry introduced a number of high-power four-wheel-drive (4WD) tractors. The first of these were built by smaller companies

such as M-R-S (1950s), Wagner (1955), Steiger (1958), and Big Bud (1961). International Harvester and Case joined in the fray in the 1960s. In 1965, Versatile Manufacturing Ltd., a new Canadian manufacturer headquartered in Winnipeg, Manitoba, entered the 4WD market with the D100 Versatile. The market, by then, was proven. The biggest problems faced by large four-wheel-drive manufacturers were high production costs and low sales volume.

WA-14 and WA-17

Deere & Company saw an opportunity to enter the 4WD market in the late 1960s while engineers finished developing a true John Deere four-wheel-drive tractor. In 1968, Deere came to a working arrangement with the Wagner Tractor Company of Portland, Oregon. Wagner had built hundreds of large 4WD tractors, mainly for the logging industry. Faced with a logging recession, Wagner listened when Deere made an offer to buy an agricultural version of its two largest tractors for 1969 and 1970.

The Deere/Wagner tractors were painted green and yellow and became an option for Deere dealers to supply their power-hungry customers. The choice was either a 178-horsepower WA-14 or a 220-horsepower WA-17.

After two years, Deere was ready to introduce its own four-wheel-drive model in 1970, the 146-horsepower 7020. Using Deere high-volume components already in the product lineup, the 7020 was a full-feature tractor, easily competing against equally powerful but less well-dressed products from competitors.

Model 4320

Farm customers in North America had their own versions of big muscle cars and rock bands to admire at Deere dealerships in the summer of 1970. They found, in 1971 tractors from Waterloo, a hopped-up version of the classic 4020, a more powerful turbocharged tractor, and a new John Deere–built four-wheel drive.

The 4320 soon was dubbed the "super 4020." It was a turbocharged version of Deere's most popular tractor. The six-cylinder diesel engine now pumped out 116

Often called the "super 4020," the 4320 was essentially a 4020 with a turbocharged engine and new tires. It was built for model years 1971 and 1972. This tractor is in southern Ontario. *Tony Gerber*

horsepower, enabling it to handle bigger and heavier field equipment. It weighed 700 pounds more than the 4020, thanks to heavier gears in the transmission, a bigger fuel tank, a larger radiator, and a wider hood. It also had a new tire size, 20.8x34, engineered to capture the added turbo power. In most other respects, it had the same standard equipment and options as the 4020, including power front-wheel drive.

An intercooler set apart the 4620 tractors built in Waterloo for 1971 and 1972. This fine, fully restored 4620 is in Ontario. *Owen Brooks*

Model 4620

The 4620 of 1970 took the original 4020 engine a step further, introducing the first intercooler on a farm tractor. It was the successor to the turbocharged 4520 customers had been admiring and buying for two years.

Cooling the incoming air made it denser and enabled the compressor to push even more air into the cylinders. It produced better combustion, boosting the diesel's output to 135 PTO horsepower.

Model 7020

Four-wheel drive took center stage in the John Deere tractor line for the 1970s with the introduction of the 7020, the first John Deere–built four-wheeldrive tractor since the premature 8010 and 8020 about ten years earlier. It also offered a new option—row-crop farming with a four-wheel-drive tractor for model years 1971 through 1975.

This 1974 John Deere 7020 still puts in the crop each spring. It's owned by Phil Isaak of Rivers, Manitoba. The 146-horsepower six-cylinder tractor was manufactured at Waterloo, Iowa, for model years 1971 through 1975.

The 7020 had a narrow hood for good visibility, and could have a narrow tire and wheel option. Single- and dual-wheel combinations were available with rack-and-pinion adjustments to fit 30- and 38-inch rows. The Syncro-Range transmission, axles, and closed-center hydraulics had proven themselves in other large John Deere tractors. The quik-coupler hitch could handle category 2 or category 3 implements.

The same turbocharged, intercooled, 404-cubic-inch engine used in the 4620 was the heart of the 146-horsepower 7020. The 7020 was about 1,000 pounds heavier than the 4620. The 1971 Nebraska Test showed this version of the engine had a 10-horsepower increase over the 4620 and, pulling with all four wheels, generated a phenomenal 40 percent increase in lugging power (18,726

pounds). An update came for model years 1973 to 1975: the updated 4630 engine was installed in the 7020, and a new Bosch inline fuel injection pump replaced the Roosa Master pump.

The 7020 had options for power steering, rear power brakes, and a 1,000-rpm PTO. A hi-lo transmission feature could increase the standard 8 forward speeds to 16. An aftermarket cab, fitted around the rollover structure, could be ordered, with optional air conditioning and a heater.

The 7020 was good for the company. Many of the components were already in high-volume production from the 4020 line. In most instances, dealers were also able to service the tractor with their existing supplies of components.

She's picked up some wear and tear in more than 30 years of farm and field labor, but there's still lots of power in this 6030 workhorse—if someone takes the challenge to pull it out of this pasture in southern Manitoba. These successors to the 5010 and 5020 were built at Waterloo from 1972 through 1977.

Three for 1971

Three tractors and a new slogan were introduced to consumers in the summer of 1971. Deere & Company was preparing for a sequel to D-Day in Dallas the following year, but it still put on a good display in this lead-up year.

Model 2030

The 2030 tractor from Dubuque's factory was the first 30 Series tractor in Generation II. It was the smallest of the three introductions and a real iron workhorse.

Power output for the improved four-cylinder engine was boosted to 60.6 PTO horsepower in the Nebraska Test in 1971. It was only 100 pounds heavier than the

2020, but it out-pulled its predecessor by 930 pounds. It had a constant-mesh transmission with options for hi-lo or a direction reverser. It also carried the proven features of the earlier 2020 model. List retail price at the end of production was $6,235.

Model 6030

A second 30 Series Generation II tractor among the 1971 introductions was the 175-horsepower 6030 from Waterloo. It was a big-muscle successor to the standard 5010 introduced in 1963 and the 5020 of 1966.

The 6030 was the largest John Deere two-wheel-drive tractor, intended for large farms growing small grains on

The Waterloo factory built this massive and powerful 7520 four-wheel-drive tractor from 1972 through 1975. The 22,320-pound machine cranked out 176 horsepower on the PTO with its six-cylinder diesel engine. The fuel tank carried 110 gallons.

the Great Plains and prairies. It could be equipped with duals and was big, comfortable, and powerful. With ballast, the tractor weighed 18,180 pounds. It did not have a factory-installed cab, but one could be ordered. The last list price was $28,745.

Under the hood, the 6030 had a much more powerful engine than its 5020 predecessor from six years earlier. The classic 404-cubic-inch engine, first seen in the 4020, had been given bigger cylinders. It now had a 4.75x5.00-inch bore and stroke, increasing displacement to 531 cubic inches. It also had a valve-in-head, variable-speed engine. It was both turbocharged and intercooled, like its predecessor. On the PTO, it generated 176 horsepower. Pulling power was tested at 15,300 pounds, and the hitch could lift 5,900 pounds.

Deere offered a unique engine option a year after the 6030's introduction: customers could order it without a turbocharger. This version came with the familiar 133-horsepower engine that had powered the 5020.

Forty-five units were built without the turbocharged engine; 3,983 were equipped with the turbocharger.

Model 7520

The 176-horsepower 7520 tractor from Waterloo, the last of the New Generation 20 Series, was nearly identical to the 7020 introduced two years earlier. It was ready for row-crop farming, as well as grain farming, with a four-wheel-drive approach.

The 7520 was equipped with the same 176-horsepower, six-cylinder, turbocharged, and intercooled diesel engine used in the 6030 tractor. Extra traction gave it a maximum pull of 22,300 pounds. It weighed about 4,000 pounds more than the 6030, and its last list price was $25,189.

The 7520 was an immediate success. More than 3,900 were sold in the U.S. and Canada in the three production years. For its market, the 6030 was also very successful and produced sales of about 2,580 tractors.

While war protesters were marching in 1968, business was carrying on. This photo, taken June 1968, shows Bill Hewitt between Hovsys Hovsepian and Michael Frank during a visit in Tehran, Iran. The other two individuals were with Deere & Company's marketing division in Mannheim, Germany. *Anna Wolfe collection*

Late 1960s Ventures

In the 1960s, John Deere tractors became competitive in tractor markets across South America, Central America, Western Europe, southern Africa, and Australia.

In 1962, the company had purchased a majority interest in South African Cultivators, a farm implement firm near Johannesburg. For most of the decade, the Deere-owned factories in Mexico, Argentina, Spain, and South Africa remained separate suppliers of niche markets. Elements of the Worldwide concept were integrated whenever possible, but each factory or country had challenges that kept the operations separate from the Worldwide mainstream.

Ten years later, after a decade of trouble and red ink, the protracted struggle with getting firm footing as a multinational corporation was coming to an end. Deere

investments overseas amounted to more than $250 million in late 1970. Half of that had been in Germany, with other significant investments in France ($46.8 million), Argentina ($39.3 million), Spain ($15.1 million), Mexico ($8.4 million), South Africa ($5.4 million), and Australia ($4.5 million).

Deere products had faced cutthroat competitors over the years, as well as bad press in Europe about American power and the Vietnam War. For a while, Deere managers considered the feasibility of merging international operations with a strong foreign company; two were considered.

Extended negotiations about a possible merger with Germany's Klockner-Humboldt-Deutz AG came to an end in 1970. The manufacturers were similar in size, but the product lines were quite different. Hewitt decided that the chemistry didn't work and broke off negotiations

A handful of individuals have begun to restore and collect the John Deere New Generation tractors. This series has been assembled by Owen and Lynn Brooks of New Lowell, Ontario. Shown in this photo, left to right, are the 2520, 3020, 4000, 4020, 4320, and 4620. *Owen Brooks*

with Deutz, immediately launching a second effort with the Italian manufacturer Fiat. This venture seemed close for a while, but ended in January 1972 with a board of directors vote against the Deere-Fiat merger. By then the board had decided that Deere could survive in Europe without a corporate partner.

In 1970, John Deere products could also be found in the Middle East, Turkey, and Iran. In 1969, Deere had invested about $200,000 in the construction of a combine assembly factory in Tarsus, Turkey, in partnership with a dealer organization that had sold Deere and Caterpillar equipment for more than 20 years. The Tarsus facility was the only combine factory in Turkey and produced about 400 combines a year. Deere sold its interest in the factory in 1979 due to the changing political situation in Turkey.

The venture in Iran was less successful. Deere had accepted a government invitation to set up a joint venture company in the town of Arak in the late 1960s for the partial manufacture, assembly, and sale of Deere

combines. The company had also provided components to Deere's combine factory in Zweibrucken, Germany. The joint effort had worked for a while, but was suspended during the Iranian hostage crisis.

Management structure and engineering at Deere also underwent serious reorganization, in 1970. Reflecting its growing diversification, the company restructured into three operating divisions: Farm Equipment and Consumer Products, U.S. and Canada; Farm Equipment and Consumer Products, Overseas; and Industrial Equipment, with worldwide responsibilities.

Engineering research, until 1970, had remained a vested interest of individual factory managers and their engineering groups. That changed in August 1970 after the board moved to replace the decentralized research efforts with an integrated structure. Two committees were established: an engineering council and a product engineering technical committee. The combination would provide corporate coordination and organized communication throughout the worldwide company.

Four-wheel-drive tractors came of age in the mid-1970s. The John Deere 8630 cranked the heads of neighbors as they watched it operating with big equipment and big comfort. More than 30 years later, many still perform the heavy field operations. This isn't a collector's item; this is a working tractor for big farms.

GENERATION II: 1972–1977

Fully Engaged Boomers

By 1972, Bill Hewitt had been head of Deere & Company for 17 years. During his tenure, the world had changed dramatically. Millions of Baby Boomers had finished high school and college. As a parent of three and a leader of a giant corporation for more than a decade and a half, Bill Hewitt began to make end-game plans. At a meeting in 1973, with Hewitt's recommendation, the 15-member board approved its first outside member, and appointed another outsider as Bill Hewitt's likely successor.

The wars for civil rights and against the Vietnam War had been won. The Cold War wasn't as cold or frightful anymore. This generation was better educated and wealthier than the previous one. They were becoming socially sensitive, and were developing their own values about how and what the world should be. They had seen the world, were on their way somewhere, and they were fully engaged in the Global Village.

When these veterans of both domestic and foreign wars returned to their farms to raise families, they came with a vision reflecting their experiences. Most were looking for modern equipment. They wanted larger machines, comfort, new technology, easy communication, and reliability in farm and business relationships.

The farm machinery industry had matured over the years. Production of tractors in the United States had peaked in 1951 at 564,000. The number of tractors on farms had peaked in 1965 at 4,787,000. Since then, both numbers had been slowly declining. Total tractor horsepower, on the other hand, continued to rise. The average American farm in 1975 had twice as much horsepower as the farm of 1960.

There were also fewer competitors among major tractor manufacturers. By 1970, the only serious competitors were Deere & Company, Massey-Ferguson, and

In 1974, while Deere & Company was setting new standards for farming, the company's leader was also at the biggest trade table in the world. Here in the White House, President Gerald Ford (right, center) has Secretary of State Henry Kissinger at his side and is facing William Alexander Hewitt, chairman, National Council for U.S.-China Trade. A Chinese trade delegation is also at the table. These discussions led to China's return to international trading. *Anna Wolfe collection*

Tish and Bill Hewitt with sculptor Henry Moore at his home in Much Hadham, England. The back notation reads: "That day we decided on his 'Hill Arches' for Deere & Company." The monumental four-piece bronze sculpture was produced in 1973 and placed at the Moline headquarters in 1978. It is one of the greatest sculptures of the twentieth century. *Gilbert Lloyd*

Deere kept the name but morphed into a new business entity between 1955 and 1965. Dealers changed as well. In December 1972, Fred Haar Company left downtown Freeman, South Dakota, for the outskirts along the highway and a new, much larger building. Alfred "Al" Haar recalled it as "one of the best things we ever did. It took me three or four years of planning … [but] we made the change at the right time." *Al Haar*

International Harvester. Deere had sold approximately a third of all tractors in the U.S. market over the previous decade, and its market share exceeded 40 percent in 1966, 1970, and 1973. Farm commodity prices rose in 1973, along with equipment sales.

Deere itself had changed too. The tall ex-Californian who chaired the board of directors had transformed the midsize Midwestern tractor company into more than just the world's largest producer of agricultural equipment. Now it was also actively involved in manufacturing construction and landscaping equipment. Sales had surged from less than $300 million in 1955, when Hewitt became president, to more than $1 billion in 1966 and $2 billion in 1973.

North American farms had invested approximately $3 billion in farm equipment in 1970. Deere clearly led the industry with 26 percent of sales and an annual growth rate of 21 percent. International Harvester had 18 percent and Massey-Ferguson was third at 8 percent. Behind the

three leaders, with 4 to 6 percent of the market, were Allis-Chalmers, Ford, J. I. Case, and Sperry-New Holland.

With significant profits for shareholders and a secure place in the industry, Deere & Company could have cut back on their investments and simply maintained their status while putting larger returns into the hands of shareholders. But that wasn't the nature of the company or its leader. In the 1970s, Deere would invest nearly $2 billion in state-of-the-art manufacturing facilities.

Hewitt was invited, through John D. Rockefeller, to join the Trilateral Commission, a new organization of elite world leaders that included his old friend Robert McNamara. The Trilateral Commission was a private organization, established to foster closer cooperation between the United States, Europe, and Japan. It was founded in July 1973 by David Rockefeller, who was chairman of the Council on Foreign Relations at that time. Later that same year, when it was time to re-open trade with the People's Republic of China, the 10-member trade mission included

Bill Hewitt and Gabriel Hauge, his old cabinmate from the navy. Hewitt's venue had changed again, from naval commander to corporate command to consultant in the hallways of world leadership.

Generation II Tractors

Within Deere & Company, Hewitt had directed the engineering department to continually improve safety and comfort, as well as power and operating features, after the fleet of New Generation tractors was introduced in 1960. By 1972, at least six years of development had gone into the project that became known as Generation II tractors for Deere's North American operations. Soon after these were released, European dealers were given an opportunity to view attractive new models from Mannheim, Germany. Tractors were also being revised and readied for introduction at Deere factories in Spain, Mexico, Argentina, and Australia.

At an open house event on August 19, 1972, in Germany, four completely new and restyled medium-size John Deere row-crop tractors for North America were revealed. They were the first of the Generation II tractor group, and were also known as the 30 Series. They were immediately, obviously different from anything seen before. This technology put the company a generation ahead of its competitors as it introduced the first true John Deere cab with a Sound-Gard body.

Sound-Gard was the successor to the Roll-Gard structure introduced in 1966. Roll-Gard had been a great idea that just wasn't accepted. Deere had wanted to make tractors safer, and the structure had achieved this; but it added $500 or more to the cost of the tractor, and farmers weren't buying it.

Bill Hewitt, seeing the response, took two strategic steps. First, he literally gave away the Roll-Gard patents, and millions spent in research, to any competitor who

Generation II "Sound-Idea" tractors, like this 80-horsepower 4030, introduced in 1972, were an immediate success for Deere and for dealers. A little rust shows on the loader for this backwoods, hardworking, 35-plus-year-old 4030 in Canada, but the loader doesn't have John Deere paint. It lives at Ron Kostenchuk's place near Riding Mountain, Manitoba.

would agree to make roll bars standard equipment on tractors. Second, he ordered the development of a tractor cab that would fully integrate rollover protection, along with comforts and safety features matching or better than those found in American cars and trucks.

Working with Deere's engineers, Henry Dreyfuss and Associates partner Jim Connor directed the Sound-Gard project. Commenting to author Randy Leffingwell years later, he recalled:

> *It was decided that we were going to do the best cab possible, that it was going to have all the safety features, all the human factors features that a cab could have: rollover protection; sound insulation; heat insulation; operator insulation from vibration and dust; air conditioning; pressurization so the air goes out, does not get sucked in; improved vision; easy entry and exit; flat platform; the second generation of the (Janet) Travell-designed seat. The cab on the Generation II tractors just blew everybody away.*

Along with Sound-Gard, a second aspect of the restyled tractors garnered favor among the gathered dealers: a new hood style. Connor explains:

> *The down-sloping hood which came out on the 1972 tractors, from a design standpoint had a couple things going for it. One, to let it slope downward gave you better visibility over the nose. Two, it gave you more space under the back of the hood to get stuff in— where you always have more trouble and need more space, because there is more stuff that needs to be in there. So we brought the back up, front down, and curved the top. Now, there was no longer a horizontal horizon line to look at, to sight along. It picked up what we started with the tractors in 1960.*

The new "Sound-Idea" tractors from Waterloo in 1972 were the 80-horsepower 4030, 100-horsepower 4230, 125-horsepower 4430, and 150-horsepower 4630. The optional Sound-Gard body was exceptionally quiet. The

The 4230 replaced the 4020 and could be purchased with either a power shift or 16-speed Quad-Range transmission. This standard 4230 diesel Generation II unit is restored and in the collection of Brent and Gregg Campbell of Brandon, Manitoba.

cab was a self-contained module mounted on the tractor chassis and isolated by four rubber mounts. The seat, the floor, and the controls were vibration-free, and rollover protection was built in. Buyers could choose a four-post Roll-Gard structure as an option to the cab, but most tractors left the factory with the Sound-Gard cab.

Among the Sound-Idea tractors, the most popular was the 4430. Deere sold 74,580 of these tractors between 1972 and 1977, and the replacement 4440 continued selling at the same pace from 1978 to 1980. It is estimated that three out of four 4430 tractors were sold with the Sound-Gard cab. The four medium-size Waterloo row-crop tractors stayed in production for six years.

The groundbreaking cab design was pressurized to keep out dust and dirt. It had side and rear windows that could swing out, unless the buyer bought the optional air conditioning feature. A heater was also optional. The windshield was large and curved with twin wiper blades, and the rear window had a wiper as well. All the cab's glass was tinted and polarized.

In addition to the new cab and hood, the four models came with a Syncro-Range transmission as

The 4430 diesel was the most popular of the Generation II tractors. It was built at Waterloo for model years 1973 through 1977. It had a turbocharged, liquid-cooled, 404-cubic-inch engine generating 125 horsepower and was sold in the United States and Canada. This one in Manitoba is still serving.

standard equipment. The transmission was coupled with a very durable, hydraulic, wet-type clutch dubbed the Perma-Clutch. It used circulating oil to dissipate heat. The 4030, 4230, and 4430 could be purchased with an optional 16-speed Quad-Range transmission, a blend of the Syncro-Range and a built-in hi-lo no-clutch shift. The three larger models were also available with power shift.

For farmers needing more power, a pair of new articulated four-wheel-drive tractors would soon be revealed.

The 80-horsepower Model 4030 replaced the popular 3020 New Generation tractor. It had an extra nine horsepower generated by a six-cylinder, 300-series gas or diesel engine.

The 100-horsepower 4230 replaced the 4020 tractor. It could be purchased with the same 404-cubic-inch gas or diesel engine, or with an optional 362-cubic-inch gas engine. It continued to come in four styles: row-crop, standard, low-profile, or hi-crop. The end of production list price was $20,432.

The 125-horsepower 4430 took the place of the 4320 in the dealer's line. It was powered by a turbocharged 404-cubic-inch engine, which made it a very different and

Another view of the 4430 diesel standard, which is still one of the best-looking tractors found on farms. Notice the narrow posts and wide expanse of glass for great all-round vision.

This unusual 4630 was waiting for a makeover at Steve's Tractor, a tractor restoration shop in Firdale, Manitoba. The front axle has hydraulic front-wheel drive, a feature common in Europe but seldom seen in Canada.

very popular tractor. Sales reports for 1975 showed dealers selling 13 of these 4430 tractors for every 5 of the 4230 model. Dealers in Canada and the U.S. that year sold 13,782 of this model tractor, nearly as many as the other three models combined. The end of production list price was $22,451.

The 150-horsepower 4630 replaced the turbocharged 4620 tractor. In the 4630, power came from a turbocharged and intercooled version of the classic 404-cubic-inch six-cylinder engine. The end of production list price was $27,076, twice the price of the 4620 at the end of its production.

Four-Wheel Drive Comes of Age

After the Sound-Gard Generation II row-crop tractors were accepted in the market, it was time for a similar update to the successful high-horsepower, big farm tractors.

In 1974, the 141-horsepower 7020 and the 175-horsepower 7520 were replaced by the 8430 and 8630, a pair

of ultimate-size, articulated, four-wheel-drive tractors with new Sound-Gard bodies and new styling. They were ideal for the largest row-crop operations as well as the largest grain farm. Both tractors had big, new, turbocharged and intercooled six-cylinder diesel engines. Both came with sixteen-speed Quad-Range constant-mesh transmissions. Perma-Clutch, power brakes, and closed-center hydraulics were standard.

The tractors didn't require new implements, but they could handle anything available. It became common to put dual wheels on them for handling larger implements and tougher field conditions. A hydraulic Quik-Coupler for a new category 3-N hitch allowed connections to both category 2 and category 3 implements. Both tractors had up to three remote cylinder outlets and independent 1,000-rpm PTO for the flexibility expected by Baby Boomers and their families.

They were built for four model years, through 1978. The last list price for the larger model was $52,475.

John Deere 8630 articulated four-wheel-drive tractors were built at Waterloo for model years 1975 through 1978. This tractor weighs about 25,000 pounds without duals, produces 225 horsepower, and still puts in the crop at Olmstead Farms of Brookdale, Manitoba.

Model 8430

With a 466-cubic-inch engine, the John Deere 8430 produced 178 PTO horsepower. This engine went on to become one of Deere & Company's most popular. Between 1975 and 1990, it was installed in 19 different tractor models and in Titan combines. Most of the engines were installed with a turbocharger and aftercooler, but some were naturally aspirated. PTO power output varied, according to the Nebraska Tests, from a low of 101.5 horsepower on the 1983 Model 4050 tractor to a maximum of 202.7 horsepower on the 1989 Model 4955 tractor.

Model 8630

The 8630 carried an engine with a huge 619-cubic-inch displacement. It produced 225 PTO horsepower and 202 drawbar horsepower. The 8630 reached maximum horsepower at 4.6 mph, with a pulling capacity of 26,443 pounds. In the 1975 Nebraska Test, it weighed 26,480 pounds with ballast.

Chairman Bill Hewitt was moving in the leagues of presidents, kings, and prime ministers in the 1970s. In this 1973 photo, he's greeting Sweden's young King Carl Gustav and Sweden's First Marshall at Deere headquarters. The Hewitt family also visited Sweden, touring the Volvo factory with the King. *Anna Wolfe*

Worldwide Utility Models

While Generation II tractors dominated tractor sales in North America for the next few years, Deere & Company continued to produce fresh and innovative tractors in the utility-size Worldwide line. In a detailed listing of Worldwide tractors from 1972–1982, Deere identifies 77 separate tractors by country of production and model number. These tractors arrived at dealerships in two waves.

Deere first introduced a 30 Series set of models in 1974, and two years later followed up with a 40 Series upgrade. These Worldwide tractors were manufactured and assembled in factories from Dubuque to Germany, Spain, Mexico, Argentina, and even Australia between 1973 and 1983. Dealers in Spain and Argentina had five sizes and models, each differently numbered, in the 30 Series Worldwide line; Mexico and Australia each had four models. There were more models in the second wave.

Tractors in the 30 Series from Dubuque and Mannheim generally had three or four cylinders and up to 66 horsepower. The largest offshore tractors in the Worldwide group had six cylinders and more than 100 horsepower. For the U.S. 40 Series, the PTO power output was defined more neatly: the three-cylinder utility tractors offered 40 or 50 horsepower; the four-cylinder tractors offered 60 or 70 horsepower; and the six-cylinder tractors offered 80 to 110 horsepower.

Mannheim became home to the three-cylinder Worldwide tractors in 1973, introducing the 35-horsepower 830 (a 4-horsepower upgrade to the 820) and the 1530, which had 45 horsepower. Meanwhile, Dubuque lost the 1520, but continued building its four-cylinder 2030 and added a completely new four-cylinder tractor, the 70-horsepower 2630.

30 Series

The four 30 Series tractors retained styling already in use and were promoted as being cheaper than the competition. Base prices for the four U.S. models ranged from about $6,000 to $10,000. Annual sales volume in the U.S. was about 3,000 to 4,500 units for each model.

Two other three-cylinder tractors from Mannheim, the 920 and the 1120, were being used in Europe and in Canada. Introduced late in 1973, the upgraded versions that replaced these two were greatly improved

This little four-cylinder 2130 diesel utility tractor could be nearing its sunset years but still gets lots of work at the Norman Smith farm near Plumas, Manitoba. The tractor is equipped with the popular 146 loader and a set of forks for loading bales. Model years were 1974 through 1979. It was built at Mannheim, with Worldwide series components, for the Canadian market.

four-cylinder-powered tractors. To avoid confusion with Dubuque's four-cylinder tractors, the tractors were given new model numbers, 1830 and 2130. The 66-horsepower 2130 was Mannheim's top-of-the-line Worldwide tractor for Canadians. It had a 239-cubic-inch displacement and was naturally aspirated, rather than turbocharged like the Dubuque equivalent. The 1830 had new styling, produced about 60 horsepower, and was less popular. In 1975, Canadians purchased 1,903 of the larger 2130 and only 866 of the 1830.

Meanwhile, other John Deere tractors were being manufactured for Europeans. Beginning in 1973, Mannheim produced a three-cylinder 1030, as well as the four-cylinder 1830 and 2130. A six-cylinder 3130 was introduced the same year. The 3130 was made in Spain and sent to Canada, but not to the United States. Spain also manufactured a 3030 with less horsepower, for Europe only.

In 1977, Mannheim began building the six-cylinder 2840. It was nearly identical to the 3130 for Canada, but it was distributed in the United States and Canada. Production of both ended in 1979.

40 Series

A 40 Series utility tractor with five models was introduced in the United States in the summer of 1975 and early 1976. All five had Generation II styling. They didn't have the Sound-Gard cab option, but the Roll-Gard protective structure was available. They were more powerful and were released in orderly 10-horsepower power increments. They stayed in production for three to seven years.

Two three-cylinder tractors in the series were built at Mannheim Works: the 40-horsepower 2040 and 50-horsepower 2240. Each offered 5 horsepower more than its predecessor. Dubuque built the four-cylinder

The 3130 seen here is a bit of an oddball. The John Deere factory at Getafe, Spain, began building a six-cylinder 3130 for Canada in 1973. Four years later, the Mannheim factory began building the 2840 tractor, which was nearly identical to the 3130. It was available to both Canada and the United States, until production of both models ended in 1979. This one, in original condition, is at the Odanah Hutterite Colony near Minnedosa, Manitoba.

The 2040 utility tractor is helping with plot work at the new Pioneer Hi-Bred Research Station near Carman, Manitoba.

60-horsepower 2440 and the 70-horsepower 2640. A few months later, in 1976, Mannheim added a six-cylinder 80-horsepower 2840 to the line.

These tractors had many features in common with the larger row-crop models. Drivers didn't have the comforts of the air-conditioned cab, but they found most of the other features at their fingertips. They had power steering, hydraulic brakes, a live PTO, a three-point hitch with lower link sensing, a differential lock they could engage on the move, a planetary final drive, and a fully adjustable swinging drawbar for implements.

The three-cylinder 41-horsepower 2040 was popular, long-lived, and had some variations. They were manufactured at Mannheim, beginning with the 1976 models, and were sold in the United States. They had a 74.4-inch wheelbase. A second 2040, sold in the same years, was also built at Mannheim. It had four cylinders with 60 horsepower, and sold in Canada. With a 90.2-inch wheelbase, it was larger than the American 2040. In

turn, it was succeeded for model years 1981–1987 by the 75-horsepower 2040S, which used the same wheelbase.

The 50-horsepower 2240 was particularly popular. Its predecessor, the 1530, had been in production for three years. The 2240 stayed in production for seven years. The two tractors both weighed 4,500 pounds and had 74.4-inch wheelbases. The newer 2240, however, had a 5-horsepower advantage due to a 179-cubic-inch engine displacement, 15 more horsepower than the 1530's engine. The 2440's base price was about $16,000, twice the cost of the 1530.

With a narrow tread, low height, and no Roll-Gard structure or canopy, the 2240 could be set up for either vineyard or orchard work, as well as for light field work such as cutting hay. The vineyard version had narrow tires, and the orchard model could be ordered with low, wide tires. Mechanical front-wheel drive made these very maneuverable in tight settings. The rear wheels had a rack-and-pinion adjustment.

This 7000 Series John Deere planter was built between 1974 and 1982 but was being used to put in a few acres in 2009. It has the Max-Emerge disc opening system.

Other utility tractors on the market in the U.S. couldn't really compete against these full row-crop tractor features on a utility tractor. Deere introduced them with a new slogan: "Family styling that's inherited—a family reputation that's earned."

Field Equipment

Field equipment for nearly every farm need continued to be produced and improved by Deere & Company for the Generation II tractors of the 1970s. Only a few of the products are highlighted here.

Tillage

In 1975, sales of 50,522 tractors in the United States and Canada were complemented by sales of 38,543 pieces of tillage equipment. These included 14,159 sets of disks, 7,485 semi-integral moldboard plows, 4,779 field cultivators, 4,581 chisel plows, and other pieces for tillage work.

Disk sales had sharply declined since New Generation tractors were introduced. Baby Boomers gradually reduced tillage and plowing in favor of more conservative land management. In 1973, nine Level-Action discs were introduced. They were called double-offset, having two front gangs overlapping to take out the middles. They came in three families based on weight per blade. Small utility tractors were matched with a 40-pound blade; the big four-wheel-drive tractors pulled discs weighing up to 212 pounds per blade.

With more power available in the 1970s, the new 1000 drawn field cultivator led sales of the integral field cultivator. The largest 1000 was a five-section folding model that was 60 feet wide.

The 1600 rigid drawn chisel plow came in sizes from 8 to 19 feet wide; the 1600 flexible folding model came in sizes from 20 to 41 feet wide. They left a rough surface with lots of residue on top, helping fields capture blowing snow.

The V-ripper had sales of almost 1,000 units annually, mostly in the Corn Belt as a fall tillage tool to conserve soil and water while loosening compacted soil.

The mulch tiller (a combination of disk and chisel) was introduced in 1972. A front-mounted disk reduced stalk length, eliminating plugging problems with the chisel plow that followed behind it.

Planting

In 1975, sales of tractors in the United States and Canada were complemented by sales of 64,746 units for planting and crop care. These included 23,382 planter units, 9,383 planters, 12,636 grain drills, and 9,840 rotary hoes.

Row-crop planters for corn, soybeans, and cotton underwent considerable development in the 1960s and 1970s. John Deere introduced the Max-Emerge line in 1975, giving a choice of the drawn 7000 or integral 7100. Particularly for corn and beans, they gave a high level of seeding accuracy and were quickly adopted by growers. The new finger-pickup meter, particularly in corn, allowed accurate planting despite irregular seed size and shape. It allowed seed to be used without first being graded.

John Deere was making grain drills for differing working conditions in this era, with more than half carrying an end-wheel design. Most were hitched as single units, but two- and three-drill hitches were available. They could have single- or double-disk openers. The LZ-B lister drill, with two ranks of staggered hoe openers, was the most popular press-wheel drill in the 1960s and 1970s. The press wheels provided good seed-soil contact.

The 400 series rotary hoe was a major innovation and leading seller in the 1970s and 1980s. It could fit the contour of a bedded crop as well as climb over rocks and stumps at high speed without tooth damage. On crusted soil, flexible down pressure of more than 30 pounds per wheel ensured penetration of the crust.

Hay and Forage

In 1975, sales of tractors in the United States and Canada were complemented by sales of 29,987 pieces of hay and forage equipment. Sales of 7,915 square balers led the market, followed by 4,620 mowers, 4,150 mower/conditioners, 3,756 side-delivery rakes, and 3,874 forage harvesters.

Major changes occurred in John Deere self-propelled combines in the 1960s and 1970s. Compare this 4400, built between 1970 and 1979, with the preceding 95, introduced in 1965. Motoring along on the road behind this pastured 4400 is another machine, a modern John Deere self-propelled sprayer.

Not very impressive today, but this John Deere 95 self-propelled combine was the best-selling harvesting machine in the late 1960s. This one is parked in a public collection at the Manitoba Agricultural Museum and was providing shade for a visitor.

John Deere mowers were an immediate success, starting with the three-point hitch and caster-wheel mowers of the late 1950s. In 1973, a major mower redesign was introduced. One was the three-point hitch sicklebar 350 mower; the other was a trail-type 450 mower. Both featured a pitman-less drive.

Farmers quickly adopted the two-stage mower and conditioner for hay when it was introduced by John Deere. The 1209 mower/conditioner was a bold step forward in 1974. It permitted higher travel speeds and a uniform cutting height in rough hayfields.

Deere introduced two options for baling hay in the mid-1970s. The first option, the stack wagon, was offered in three sizes. Cured windrows were lifted by a flail pickup and thrown into the wagon. The hay was compressed and eventually released as a stack in the field. The second option—large round bales—was highly successful. The John Deere 500 round baler entered the market in 1975, followed by the 410 and 510 in 1977. Round bales became the preferred bale technology in the next decade.

Two self-propelled forage harvesters, the 5200 and 5400, were added to harvesting options in 1972. They offered a high-efficiency way of handling forage with dedicated, self-powered machines. A hydrostatic

Another essential item for livestock operations from the 1960s and 1970s was this John Deere square bale elevator. More than 10,000 of these 200 series bale elevators were made between 1961 and 1965.

Livestock operators depended on Deere equipment for all aspects of the farm, including removing and spreading the manure. The 33 dairy spreader was manufactured between 1961 and 1969.

rear-wheel-drive option helped growers continue harvesting in rainy weather.

Harvesting

In 1975, sales of tractors in the United States and Canada were complemented by sales of 45,328 pieces of harvesting equipment. Sales were led by 13,883 headers, followed by 12,976 self-propelled combines and 10,843 corn heads.

New Generation John Deere self-propelled combines were introduced with the 1970 harvest and updated in 1974. Harvesters had a choice of the three-walker 3300 or four-walker 4400 in one design, or a four-walker 5500 or five-walker 7700 in a second design. For all four combines, the engine was located in front of the grain tank and the cab was offset to the left of the engine. A reverse-flow engine fan pulled air through a large front screen; the cab was integrated with the design, making it quieter and less dusty. Updates in 1974 had a better engine cooling system and a rotating screen that removed trash before it could accumulate on the screen.

Important innovations in tools for cutting row crops and getting them into the combine were released in the 1970s. First came the 40 series corn head in 1970. It combined several features that engineers had been developing for 15 years and was much superior to anything on the market for corn. Low-slope gathering points helped salvage corn that was down or lodged. A shield kept ears from rolling out the front. Many were fitted to the front of competitors' combines.

Deere Harvester Works designed, tested, developed, and began producing the 50 series row-crop head in 1975. It was a flexible platform, able to harvest 1.5 bushels per acre more than a rigid platform in soybeans. It also was ideal for sorghum and sunflowers.

Farmsteads

John Deere sold 43,174 pieces of equipment for the modern farmstead in the United States and Canada in 1975. Leading these were sales of 11,758 farm loaders and 11,013 wagons. Dealers also sold 6,992 rotary cutters, 3,812 rear blades, and 3,290 spreaders.

Rotary cutters for a variety of purposes started coming into use in the 1940s. In 1972, John Deere unveiled a new line of heavy-duty rotary cutters (the 509, 609, and 709) that offered a cut of five to seven feet. They could cut two-inch brush and were a good choice for weed control in pastures as well as along roadsides.

John Deere became recognized as a major loader manufacturer in the 1950s. Three models were updated in the first half of the 1970s and remained in production

ERA II for the industrial division began in the late 1960s. One of the big sellers was this machine, the world's first articulated grader. This photo is from March 1967. *Deere & Company*

A new version of the grader, the John Deere 570-A articulated motor grader, was on the road in 1973. *Deere & Company*

into the 1990s. They are the 146, 148, and 158 loader for various sizes of tractors.

Industrial Division

Construction or industrial uses for John Deere tractors began early, before the first Model B two-cylinder tractor was painted industrial yellow in 1936. Marketing went to a higher level in 1949 when a new crawler-type Model MC machine came out of the brand-new Dubuque factory. Commitment to construction machinery strengthened again in 1956, in the United States and at Deere. Under President Dwight D. Eisenhower, the United States launched the Highway Trust Fund in 1956. It provided secure funding for a 40,000-mile, $40 billion highway construction program that would build the interstate highway system.

Hewitt and directors approved the formation of the John Deere Industrial Division in 1956. This was followed by the Industrial Marketing Division in 1957 and a new general manager at Dubuque Works. In the ensuing years, Dubuque became focused on building industrial equipment; tractors for agricultural uses became just one aspect of the factory. In retrospect, these years became known as ERA I for the John Deere Industrial

Division. Strategically, Deere's executive chose to avoid direct competition with Caterpillar. On the West Coast, most machinery dealers shared both a Caterpillar and a Deere franchise. Deere chose to build small and mid-size equipment, still an important market but more open to competition.

The 1960s and 1970s became a kind of golden era for construction equipment manufacturers. In addition to the highway system, huge fleets of construction machinery were used on giant projects in the United States and Canada. Large dams were built in both countries. The Churchill Falls power project in Labrador, Canada, began in 1968 and was followed by the 1973 launch of Quebec Hydro's massive James Bay project in remote northern Quebec. In the United States, an excavation project connecting the Tennessee River with the Gulf of Mexico started in 1972; it was never completed, but millions of cubic yards of earth were moved during the next 15 years. The Trans-Alaska pipeline, begun in the 1970s, was a fourth giant project. The pipeline project put 1,500 pieces of heavy construction equipment to work moving millions of cubic yards of earth. At least three major mining projects were launched in British Columbia, Arizona, and Quebec between 1966 and 1975.

A handshake in Amman, Jordan, between Bill Hewitt and King Hussein at the royal palace during a Mideast tour in 1975. Hewitt described the king as "always his own man when it came to advocating the best interests of Jordan. Very knowledgeable and likable, he communicated quite well." *Anna Wolfe collection*

ERA II

With more large-scale projects scheduled, and in light of the quality of the New Generation agricultural equipment launched in 1960, John Deere managers decided it was time for the company to become a top contender in the worldwide industrial market. Dubuque managers focused on the attractive niche in small and medium-size equipment. As ERA II unfolded, directors realigned company management so that Dubuque could headquarter a coordinated worldwide program and supervise the introduction of several lines of industrial and construction equipment. To further avoid a conflict of interest for dealers, marketing split most of the West Coast John Deere franchise dealers away from the Caterpillar franchises and separately opened John Deere industrial franchises.

At a world construction expo in 1969, Deere displayed the JD500-B backhoe and six other lines of products, proving that the company was committed to

the industrial equipment world. New lines in this era included the world's first articulated-frame motor grader, the 83-horsepower diesel JD570. That machine was followed by a large skidder in 1968 and the 131-horsepower JD690 excavator in 1969.

Management consolidated industrial design and manufacturing at Dubuque in January 1969 by moving its Industrial Equipment Works group from Moline into an operating division of the Dubuque Works. In ten months, all industrial product engineering had moved to Dubuque. A year later, the expanded Product Engineering Department at Dubuque had 550 personnel.

More large events followed. A 1971 dealer fly-in, comparable to D-Day in Dallas a decade earlier, was hosted at Dubuque. A total of 2,200 dealers and customers attended. In 1972, a training center at Dubuque was opened. By 1973, the John Deere Dubuque Works employed more than 6,200 workers.

ERA III

A startling five-year plan for the industrial equipment business was revealed at a May 1974 dealer show in Moline. The show unveiled 28 models of excavators, scrapers, crawlers, and motor graders. Some were still on the drawing boards and would go into production over the next five years. Many were two to three times larger than current models. The products were designed for construction, forestry, and utility work.

The announcement was unprecedented for the company and for the industry. Deere officials spoke openly with dealers about plans for 1974–1979 products. Plans included a large excavator, four more graders, a large four-wheel-drive loader, an improved scraper, and four crawlers with hydrostatic-drive transmissions. Many were available on display only as experimental models.

At the time, Chairman Bill Hewitt said more than $50 million had been spent on ERA III product research. He expected development expenditures in the next five years would exceed $125 million. A realistic goal, he said, was for Deere & Company industrial equipment sales to exceed $1 billion before 1980.

A second ERA II product was the 690-A excavator. This one was photographed, fresh out of the factory, in November 1973. *Deere & Company*

After the ERA III announcement in 1975, the company formed a worldwide industrial equipment marketing organization, John Deere Intercontinental Limited S.A. From an office in Brussels, it served customers in Europe, the Middle East, and Africa. This was followed in 1978 by a similar group serving markets in Latin America and the Far East.

More floor space was needed for the fast-growing industrial division. As ERA III was unfolding in Moline, construction was underway for a second John Deere industrial factory. Deere had purchased a 960-acre site near Davenport, Iowa, about 70 miles from Dubuque. In 1975 and after, Dubuque was able to transfer production of most front loaders, skidders, and motor graders to the new John Deere Davenport Works.

Meanwhile, rapid expansion continued at Dubuque. The Dubuque factory in 1970 had approximately 2.5 million square feet of floor space. Factory square footage rose to 4.3 million by 1975 and to 5.3 million in 1977. By then, the Dubuque Works size was equivalent to about 100 football fields. In the same period, employment increased to 7,664 in October 1974. Some products were transferred to the new Davenport factory, and a slight recession occurred in 1975, but by 1979 the single factory employed more than 8,000 workers.

Technology at the plants was continually upgraded. John Deere factories in the 1970s had the latest equipment, from computer-aided design to advanced manufacturing systems and inventory control. Facilities were built to design, test, and evaluate new industrial machines.

Industrial Equipment

Several pieces of equipment contributed to the expansion of John Deere's presence in the construction market during the 1970s.

Industrial products manufactured at Dubuque during the Generation II/ERA II period included this 544-B four-wheel-drive loader. This one, built between 1974 and 1980, still provides heavy-duty service at a metal recycling plant near Neepawa, Manitoba.

In 1971 the company introduced its first purpose-built or integral backhoe loader series. Deere had been the first full-line manufacturer to design and build backhoes and loaders for tractors. The JD310, 410, 500, and 510 combination loader and backhoes gave the industry an integrated machine. The machines were stronger, and the hydraulics were simplified. Hydraulic lines were protected; even the loader lift cylinders had been protected by mounting them inside the boom arms. The machines had closed-center hydraulics, smoother controls, more fuel economy, and a cooler operating temperature.

The JD690-B excavator was a 1973 innovation. The new excavator, an update to the 1969 JD690, had two-speed track propulsion and a crawler-type undercarriage.

In 1967, the company had set a new industry standard by introducing the world's first articulated motor grader, the JD470. A 1975 update was built at the Dubuque Works. The new JD770 was a "push-button" hydraulic-controlled motor grader that kept the articulated-frame steering and hydraulic control for the front blade.

In 1976, John Deere engineering introduced a huge improvement for crawler-dozer operators. The new JD750 crawler, followed by the JD850 a year later, improved operator comfort and machine performance. Each track now had an independent hydrostatic pump and motor. The Dual-Path drive allowed proportional power during turns, saving time and fuel. Hand levers, or a pedal option, controlled the speed of each track without steering clutches or brakes. Both on turns and on straight-ahead work, optimum speed was continuously matched to load. Operators, with less mechanical stress, could concentrate entirely on moving the load.

Deere & Company Executive Vice President Robert Hanson commented on his outlook for the industrial division in a September 26, 1977, interview with the Dubuque *Telegraph Herald* newspaper. The company, in his view at the time, was positioned to grow nearly twice as fast as the competition in industrial equipment due to the heavy investments being completed. He said, "Industrial equipment currently accounts for 14 percent of Deere's total annual sales and should increase by 25 percent by 1985."

A quick review of the decade at Dubuque Works shows that it succeeded by building a large, diverse line of industrial equipment as well as components for the Worldwide agricultural tractor lines.

Agricultural tractors coming out of Dubuque during the years 1968 to 1977 included models 760, 1520, 1530, 2030, 2440, 2630, and 2640. Industrial products manufactured in the same years included crawlers (models 550, 750, and 850), crawler-loaders (555 and 755), crawler/crawler-loaders (350-B, 350-C, 450-B, and 450-C), excavators (690, 690-A, 690-B, and 890), forklifts (380, 480-A, and 480-B), motor graders (570, 570-A, 670, and 770), backhoe loaders (500-C, 310, 310-A, 410, and 510), four-wheel-drive loaders (544, 544-A, 544-B, 544-H, 644, 644-A, 644-B, and 444), scrapers (760-A, 762, 860, and 860-A), skidders (540, 540-A, 540-B, 440-B, 440-C, 740, and 640), four-wheel-drive compactors (646, 646-B), and 13 models of wheel tractors.

By early 1975, approximately 85 percent of the new highway system was open to traffic. The rural portions had been completed. It was estimated that the last 15 percent would absorb 36 percent of the total cost and would require a great amount of new construction machinery.

Consumer Products

Deere & Company entered the consumer products market in 1963. It's estimated that, between 1966 and 1976, sales of consumer products shot up 500 percent, making this the fastest-growing segment of a fast-growing multinational corporation. During this era, Deere became a household name as well.

Lawns in the late 1960s grew in size and status. In the country and in the suburbs, homeowners suddenly wanted powered equipment for managing the yard. At the entry level, the gas-powered push mower became as common as the color television. By the 1970s, riding mowers could cover about twice the acres per hour as a push mower. Upgrades included the lawn tractor, the residential front mower, and the lawn-and-garden tractor.

At first, a small portion of the Horicon Works was set aside to build the new John Deere 110 lawn and garden tractor. The little machine had a Variator drive-in series with a three- or four-speed transmission. The variable-speed drive principle had been proven in John Deere combines, where two belts were used in tandem to transmit power and change the ground speed. The unit also had a quick-attach mowing deck. Sales quickly took off. One record indicates that Horicon built 22,280 units of the 110 by 1966 and kept going.

As consumer sales grew, more space was required. The last Horicon grain drill production was moved to the John Deere Des Moines Works at Ankeny, Iowa. The factory was renovated and, in 1969, was dedicated to manufacturing consumer products. By this point the factory had more than 630,000 square feet of floor space and about 1,000 employees. By 1977, Horicon had peaked at 1,584 employees.

Once a year, Deere executives would travel from Moline to Horicon for a product review meeting. Many products were considered. One retired engineer from Horicon, Lee Wanie, recalled discussions about building off-road motorcycles and outboard motors. Those proposals didn't get off the drawing boards, but many did—and they went beyond lawn equipment.

Custom Colors

A proposal that got off the ground and flopped was recalled by retired Horicon engineer Jack Hoffman. It was known as the Custom Color program and was introduced in 1969 on lawn and garden tractors. The machines that were introduced are known today as the Patio Series.

All the 1969 Custom Color lawn and garden tractors were initially painted Dogwood White. Each dealer had a supply of hoods and seats in a choice of four colors for the discriminating buyer. Buyers customized their tractor with a choice from Patio Red, Sunset Orange, Spruce Blue, or April Yellow. The 1970 Custom Color models had engines, fuel tanks, and some other parts painted black. Production ended with the 1971 line.

Hoffman recalled that the program was suggested by Henry Dreyfuss and Associates, shortly after its highly successful idea of providing color options for telephones, as opposed to plain black.

In the lawn care business, the 110 mower is a little giant. It was the first John Deere lawn and garden tractor. It was built at the Horicon Works, starting in 1963. Today, some are lawn ornaments; some are still mowing lawns. This one is in rural Manitoba.

Dreyfuss thought, these new garden tractors are bought by homeowners, and we should offer those in colors other than just the green-and-yellow. We made up sample tractors and had a big meeting down at Deere & Co headquarters in Moline. Our general manager wanted to try it on just the 60 lawn tractor, rather than the whole line. We had just about convinced the Dreyfuss people that we should try it on just one model first, when Bill Hewitt came walking through the area.

Henry Dreyfuss, being a friend of Bill Hewitt, said, "Bill, come over here. I'd like to show you something." Hewitt walked over and Dreyfuss gave his spiel about Custom Color tractors and how

customers would just love 'em. With hardly any effort, he changed over Bill Hewitt. Bill said something like, "Yeah, that sounds like a good idea." So we got our marching orders. That was all it took.

It turned out to be a fiasco. Dealers were getting Custom Color tractors, and they had yellow attachments in stock! They were having to re-paint attachments or do something. It wasn't simple in the factory, either, having all those different colors. Within about a year, Custom Color tractors died a sudden death. Customers didn't reject the colors, but they didn't jump on them either. An awful lot said, "I want a tractor like the ones out on the farm." Now, those are very collectible.

Horicon Works updated its lawn equipment with this John Deere 70, built between 1970 and 1974. It was rated for seven horsepower. Owners could attach a rear bagger. This one still is a very reliable machine. It was waiting to be auctioned at the James Crook farm near Neepawa, Manitoba.

Lawn Equipment

Deere's first lawn tractor, the 60, helped launch today's huge market. It was introduced in 1966. It had a more rugged frame and drivetrain than other brands already on the market. It could handle a front-mounted blade for light snow removal. (Not that the model wasn't in the garden tractor category, as it was designed for heavy pushing or for tillage.)

The Model 60 was replaced by the 70 in 1969 and by the 100 in 1974. Sales of these lawn tractor models peaked in 1970 as dealers sold 6,600 of the 70 series lawn tractors. Lawn tractor sales shot up again after two new units designed to work with a rear grass bagger were introduced in 1978. The 108 had 8 horsepower; the 111 had 11 horsepower.

Horicon's original 110 was superseded in 1967 with the 14-horsepower 140 lawn and garden tractor. This was John Deere's first hydrostatic-drive consumer tractor, and was a first for the industry. The 140 had three hydraulic circuits and a mid-mounted blade or deck.

New designs in lawn and garden tractors were introduced as the 200 series in 1974, still with the Variator drive, and the 300 series with hydrostatic drive. A hydrostatic 400 was introduced in 1975. Typical annual sales in the 1960s and 1970s were 30,000 units a year. By the late 1970s, the best-selling unit was the 17-horsepower 317 with hydrostatic drive.

More products for lawn and garden care came along in the 1970s. Deere entered the rear-engine riding mower market in 1970 with three machines: the 55, 56, and 57. The 56 led in sales, with more than 15,000 manufactured in both 1972 and 1973. The 57 was at the top of the line. The seven-horsepower 57 riding mower had a 34-inch cut, two forward speeds, and one reverse.

John Deere's first push mower hit their stores in 1970. It was built by Toro Manufacturing or by Lawn-Boy. A

A tough, multipurpose series of lawn and garden tractors was introduced for John Deere in 1974. This 10-horsepower 210 still handles a little snowblower (also from Horicon Works) and serves a family farm run by Kris Friesen near Tenby, Manitoba.

year later, true John Deere push-type lawn mowers were available. The Deere 20-inch mower had a pressed-steel deck, a Briggs & Stratton engine, and an optional side-mounted grass bagger.

The E90 electric riding mower was introduced in 1972, followed by up an update, the E96, in 1975. Electric riding mowers were silent and powered by batteries. The E96 was operated by three 12-volt batteries. One full charge provided power to operate for about an hour and cut roughly an acre of grass.

Another 1975 update was the 68, an eight-horsepower riding mower. It was gas-powered but nearly silent because the engine was completely enclosed. Deere sold around 16,000 units of the 68 in 1978.

Three series of lawn and garden tractors—the 200, 300, and 400—were introduced throughout 1974 and 1975 as Deere moved to replace the highly successful John Deere 100 series. The 1974 200 series (200, 210, 212, 214) kept the proven Variator drive used in the 110

tractors. The models had a variable-speed V-belt drive. The 300 and 400 series machines used a hydrostatic drive and had more power. The 210 was the best-selling model for a while, followed by the 17-horsepower hydrostatic 317 in the late 1970s.

Accessories for the lawn and garden tractors—such as mowing decks, blades, snow blowers, and garden tillers—were built at Horicon and available from dealers. Nearly every unit sold with a mowing deck; many had a second or third accessory that extended the versatility and operating season. One of the favorites for two decades was the two-wheel dump cart that attached to the rear hitch.

Horicon Works built its 500,000th lawn and garden tractor in 1970, and passed the one million milestone 14 years after building the first machine in 1963. Sizes progressed from 7 horsepower to 18 horsepower. The machines were widely used by both residential and institutional customers.

John Deere lawn and garden tractors brochure, featuring the 400 series.

Garden Machines

Building push-type garden tillers was another natural step for the world's leading tractor manufacturer. Deere introduced three sizes of front-tine garden tillers in the 1970s. The 3.5-horsepower 324 tiller came out of Horicon in 1971. This 24-inch garden tiller had a reverse gear, allowing operators to back out of tight spots without heavy lifting. The most popular was the 24-inch six-horsepower 624 garden tiller introduced in 1974. Sales exceeded 12,000 units in both 1974 and 1975. The smallest, introduced in 1978, was the two-horsepower 16-inch-wide 216 tiller. It was a handy machine for a small flower or vegetable garden.

Bicycles

Consumer demand for adult bicycles spiked in the early 1970s. Starting with a Taiwanese bicycle, engineers at Horicon developed a model for North America. The project lasted three years.

A bicycle from Deere & Company, 1974.

The Taiwan-built bicycle was introduced in 1974. It became the most popular consumer item from John Deere in 1975, with sales reaching more than 59,000. At the end of the 1976 season, it was removed from the product line due to quality issues.

Retired Horicon engineer Jack Hoffman recalled this anecdote from 1974:

97

At the time we introduced the bicycle, we introduced the advertising slogan, "Nothing runs like a Deere." It was introduced for snowmobiles originally but was picked up for other things. Our first shipment of Taiwanese bicycles came in a container. All the dignitaries from the factory and some from Deere & Co were there to see the opening. The bicycles came in skinny, long, tall boxes. When they opened the container, all these skinny boxes were inside. On the end, they were supposed to say "Nothing runs like a Deere." But the Taiwanese had decided that was too much to put on that skinny cardboard box, so everything said, "Nothing runs Deere!"

Snow-Clearing Machines

In 1970, Deere introduced three walk-behind snow blowers (the 526, 726, and 732). Model numbers represented first the horsepower and then the cutting width—for example, the 526 had a 5-horsepower engine and 26-inch cut in the snow.

Snow blowers have two stages. An auger picks up the snow, and a blower discharges it. Blowers have more power and work in deeper snow than throwers. They also move snow farther than snow throwers. Most snow blowers in the 1970s were less than 26 inches wide. Larger models, up to the 832, were introduced later in the decade. The 8-horsepower, 26-inch 826 topped snow blower sales for Deere, with dealers selling about 7,500 of this model in 1979.

Snow throwers soon followed, and became better sellers for dealers. They have a single stage rotor that both picks up the snow and throws it. About two-thirds of the units sold in the 1970s were snow throwers. The most popular unit, and the smallest, was the 3-horsepower 320 snow thrower. Dealers sold nearly 16,000 of these in 1979.

Snowmobiles

Snowmobiling had been discovered by consumers and major manufacturers in the 1960s. Skiing could be a lot more fun if the skis were mounted under a powerful little engine that would drive a wide rubber track across the

The Trailfire snowmobile was bred for agility and stability, with a 15-inch track width, according to this John Deere brochure from 1978.

snow while holding a seated operator.

In 1971, Deere entered the fast-growing recreational market with two snowmobiles from the Horicon Works and a new slogan. For the hot new 1972 John Deere 400 and 500 snowmobiles, the slogan was: "Nothing Runs Like a Deere."

Deere built 7,310 of the John Deere 400 snowmobile and 4,615 of the 500s for the introduction. John Deere snowmobiles used two-cylinder, two-cycle, high-rpm engines to achieve both high power and light weight. On the first model, the 400, the engine was mid-mounted between the operator's legs. A variable-speed V-belt drive connected the engine to the track. The track had bogie-wheel suspension. After that, new models were introduced every year.

The JDX4 and JDX8 snowmobiles became available in 1972. The JDX8 Blitz black snowmobiles offered a 440cc 42-horsepower engine at the front. They were followed by the JDX6 and 300 in 1973.

Three John Deere sleds were built to meet the regulations of a hot new sport: cross-country snowmobile racing. Minimum production was 1,000 machines, and engine size was limited. The first was the 295/S for the 1973–1974 season. It was followed by 1,000 of the 340/S, a machine with 340cc displacement. The last was the 340 Liquidator, and only 600 were required to meet the competition rules. It won the grand prize of snowmobile racing in early 1975, the 500-mile race between Minneapolis and Winnipeg. The Liquidator was similar to the 340/S, but had a liquid-cooled engine with a big radiator and head scoop. It used flying snow to cool a heat exchanger under the driver's seat. Some other exotic names in the series of Deere snowmobiles included Spitfire, Cyclone, Liquifire, Trailfire, Sportfire, Sprintfire, and Snowfire.

After building about 225,000 snow machines in many models over 13 years, in 1984, the John Deere snowmobile line was taken over by Polaris Industries of Minneapolis. For Deere, the market had proven highly unpredictable (depending on snowfall) and less than profitable despite significant investment in engineering and manufacturing.

Sales of some consumer products in 1975 included: lawn and garden tractors, 35,661; walk-behind mowers, 21,666; riding mowers, 21,294; snowmobiles, 20,612; walk-behind tillers, 16,026; and chain saws, 14,792.

In the mid-1970s, John Deere solidified its place as an industry leader with a combination of intelligent, evolutionary innovations in the agricultural line, growth in the industrial segment, and expansion into consumer products. The company proved that it could make smart moves in the relatively prosperous era of the early and mid-1970s. The times ahead of them would challenge the company to continue to lead in more challenging economic conditions.

This John Deere utility tractor, looking into the sunset near Plumas, Manitoba, was built in Germany in the 1970s. With good parts availability, it still could see many years of work before going to a collector for restoration.

The 155-horsepower 4640 could be purchased as a row-crop or standard tractor. Set up with power front-wheel drive and narrow rear duals, it could straddle rows and pull harvesting equipment as well as tillage or seeding equipment. This 1981 unit is owned by Kendel Friesen, Birnie, Manitoba.

THE IRON HORSES: 1978–1982

Galloping into Recession

The early 1980s proved to be a tough time for the agricultural economy in North America. The bright promise of agriculture seen from 1973 onward quickly faded in the 1980s. America's farmers produced bumper crops in 1980 and 1981, but their efforts resulted in large carryovers and lower grain prices. This coincided with a general recession in the world's economy that began in 1979, which seriously affected the agricultural equipment industry.

Farm production was used as a political weapon in 1980. A grain embargo was imposed by the United States after the Soviet Union invaded Afghanistan. Other exporting countries stepped into the breach to fill Soviet orders, spurring farm growth in Brazil and Argentina in particular. It also damaged the credibility of the U.S. as a reliable supplier.

Operating a large or growing farm in the 1980s required the skills of a small-to-midsize business manager. Many farmers took college or university courses to hone the skills they needed. Family members needed to help by operating a second tractor or driving the supply truck during the busier seasons.

These forward-looking farmers began reducing tillage to conserve moisture and soil. They moved into

continuous cropping and diversified crop rotations. Field speeds increased, equipment became wider, and farm chemicals and manufactured fertilizers became essential. Farm managers began using accountants and consultants.

President Jimmy Carter visited the Deere Administrative Center in 1980 while on a vacation cruise along the Mississippi River with his wife and daughter. Bill Hewitt noted he was impressed with Carter's intelligence and good nature, but felt he "would have been a more effective president if he had not tried to master all the details of the countless problems." *Anna Wolfe collection*

The six-cylinder, 110-horsepower 4240 could be purchased for about $40,000 in 1983. It weighed a bit less than the Model R two-cylinder diesel of 1950, but kicked out more than twice the horsepower while offering a quiet, comfortable cab and many more options. This 4240, owned by Gordon Gilchrist, is equipped with the 158 loader complemented by a silage grapple and materials buck with tines.

In most areas, the venerable plow that had founded a company for John Deere 140 years earlier was scrapped or turned into a yard ornament.

These and other developments put pressure on agricultural machinery manufacturers. Their customers were looking for more fuel-efficient equipment and for redesigned machinery adapted to new conservation farming practices.

North American Tractors

Deere & Company was positioned for at least some of the changes taking place. The John Deere 40 Series tractors were introduced in the fall of 1977. The Waterloo Works had five models for the big tractor market—the 4040, 4240, 4440, 4640, and the all-new 4840. The Dubuque Works had five new 40 Series utility tractors.

Updates were offered in the fall of 1979 for the Dubuque Works tractors. At the same time, Deere was preparing new tractors for the European market to be released in the 1980 season.

Deere & Company soon offered a tractor with the proper horsepower and configuration for virtually any agricultural need. By 1981, John Deere had the broadest range of tractor sizes and types on the market. A

customer in the United States could choose from three compact utility tractors starting at 22 horsepower, five utility tractors starting at 40 horsepower, five row-crop tractors starting at 90 horsepower, and two four-wheel-drive models with 180 or 228 horsepower. Topping it off, in the fall of 1981, Deere released an impressive new series in the large four-wheel-drive tractor group.

40 Series

A new set of five iron horses came galloping out of the Waterloo Works factory and into John Deere showrooms late in 1977, at about the time that an updated Product Engineering Center was being commissioned in Waterloo.

There was more iron than ever in these workhorses, and more power. They were the culmination of more than 20 years of leadership by Bill Hewitt, during which time Deere & Company had become nearly the undisputed world leader in farm machinery. The performance of these tractors in the field, and the solid customer acceptance, helped Deere to clinch that leadership position in the recession that followed.

The line of new 40 Series tractors were dubbed "Iron Horses" by John Deere's marketing department. The 40

Ten new tractors were introduced by Deere & Company in late 1977. The five from Waterloo Works were the largest and dubbed by the marketing campaign as the "Iron Horses." The 4640 was turbocharged, intercooled, and the six-cylinder 445-cubic-inch diesel engine produced 156 horsepower.

followed numerically in the progression from 10 Series (New Generation) to 20 Series (Classic) to 30 Series (Generation II). Four of the Iron Horses used the 16-speed Quad-Range transmission. They had bigger radiators and water pumps, larger fans, and higher-capacity alternators. They had bigger fuel tanks, giving the operator 10 hours to get his work done without refueling. Increased hydraulic capacity improved the performance of the hitch, loader, and remote cylinders.

Model 4040

The smallest of the new 40 Series, the 4040, carried the proven 90 PTO horsepower, 404-cubic-inch, naturally aspirated engine that traced back to the 1963 Classic 4020 tractor. The 4040 could be purchased for about $36,000 and came with a Sound-Gard cab—a comfortable private office on wheels.

Imagine the change for a seasoned operator upgrading from a 1960 Model 4010. The 84-horsepower 4010 had been the most powerful of the New Generation tractors, and a no-frills version could be purchased for less than $5,000. Weather-hardened operators faced whatever nature tossed their way on the 4010, happy for a more comfortable tractor seat and the confidence that an easier life awaited the next generation.

For the farmer needing a little more power, the next four Iron Horses just got better. The 4240 turned out 110 horsepower. The turbocharged 4440 delivered 130 horsepower. The larger, turbocharged, and intercooled 4640 had 155 horsepower. These four were available as standard and row-crop tractors. Two, the 4240 and the 4440, could be purchased as hi-crop tractors with a standard four-post Roll-Gard structure.

Model 4840

The largest in the Iron Horse family was the 4840. It came with the Sound-Gard body and the same six-cylinder 466-cubic inch 180-horsepower diesel engine as used in the articulated four-wheel-drive 8430. It was the only Iron Horse offered with the eight-speed power-shift transmission as standard equipment. It weighed 20,370 pounds with ballast in the Nebraska Test, about 2,000 pounds more than the 4640.

In short, the acres-per-day rate and the working environment for tractor operators was a generation away from the first New Generation tractors. Within the Iron Horse family, the turbocharged 4440 became the best-selling model. The heavier 4640 could be set up with power front-wheel-drive and narrow rear duals for straddling sugarbeet rows during a muddy harvest.

Deere's readiness for the changing market resulted from good, forward-looking management. Deere established an 11-member engineering task force in 1976. The team was given a year to evaluate both Deere and competitive tractors with a view toward meeting emerging farm demands in the next 25 years in the United States and Europe.

They first spent almost four months in a remote area of Texas with a set of 16 tractors, including one from Australia and another from South America. They operated, serviced, and fully evaluated the tractors, gaining direct knowledge of their strengths and weaknesses. The same team then spent four months testing 15 European tractors at a German base near the Black Forest. They traveled widely in Europe as well, interviewing farmers and observing farming practices.

"We actually did farming for German farmers, operating under their conditions and determining what needed to be provided," said Bob Haight, a member of that team. "Members from our German factory were part of the task force. In Germany, for example, whenever it would rain you were stuck if you didn't have mechanical front-wheel drive; that's an absolute necessity there, and that point comes home very rapidly when you're the one driving the tractor.

"Out of that task force a lot of things in tractors today were developed to make it a worldwide application. Bill Hewitt had a vision that he wanted a worldwide tractor and you had to know what it would take. That's what we were trying to find out."

In principle, the task force tried to develop a single tractor that would meet the differing conditions and laws of many European countries. Europe's small, irregular, and sloping fields required a high level of maneuverability. At the time, front-wheel-assist tractors were difficult to steer. The task force recommendations helped create a highly maneuverable mechanical front-wheel-drive tractor (MFWD) design that was both agile and easy to steer. This patented design, Caster/Action, became an industry standard with the 1982 introduction of 50 Series John Deere tractors.

Hood styling seen today is more than a matter of good looks. Haight explained, "The hood drop-off, what we called the 'droop-snoop,' is because German law required that operators be able to see the ground within 12 meters all the way around, from the center of the operator. Tractors today have that real drop-off on the hood, so they can be marketed in Germany. You started seeing that influence in Deere tractors in the 1980s. Now, it's very pronounced in tractors."

Four-Wheel Drive

Following on the hooves of the rear-wheel-drive Iron Horses, Deere & Company replaced its two largest four-wheel-drive tractors with new models for 1979. The chassis was even larger, able to handle the largest pieces of equipment for deep plowing or tillage.

The 8440 and 8640 had a bit more horsepower and offered an optional front-and-rear hydraulic differential lock, four remote cylinder outlets, a new single-lever Quad-Range transmission control, and a new Hydra-Cushion seat promising more operator comfort. These Waterloo-built 40 Series four-wheel-drive tractors were designed to operate at full power continually with heavy field equipment where traction wasn't an issue (usually above third gear).

Both tractors now offered a self-monitoring system called Investigator. It monitored many functions in the engine and powertrain and promised to warn operators about emerging issues.

Utility Updates

For the 1980 season, four of the five 40 Series midsize utility tractors from 1975–1976 were updated to Generation II styling without a model number change. However,

Deere introduced two big four-wheel-drive tractors in 1979, the 8440 and 8640. Both had a class III hitch ready to lift more than 8,000 pounds. They looked a lot like the predecessor but had many improvements for function and durability. They were built for model years 1979 through 1982. This 8440 still works for Doug Dowsett in Clanwilliam, Manitoba.

the 80-horsepower 2840 was renumbered; it became the 2940 for 1980 and had the advanced styling of others in the series.

These advanced tractors were marketed differently in the United States and Canada. For the United States, the updated series was marketed as "New Profiles of Performance." For Canada, the same 40 Series tractors were called "Schedule Masters."

Confident with engineering improvements made in the past four years, Deere broke ground by offering buyers a longer warranty. Warranty coverage was increased to 1,500 hours or two years for the engine and transmission. As a group, the updated utility tractors also had more comfortable seating, an electronic instrument panel, and a new parking brake.

These tractors had a distinctly different look than their predecessors. The 40 Series utility tractors had a new profile. They had the higher, down-sloping hood of the Generation II row-crop tractors and a remodeled dash with an electronic instrument cluster. This enabled Deere to offer an optional Sound-Gard cab for quiet, safe, and more comfortable year-round farm operation.

The new utility machines also had performance changes. Cylinder bore was increased for two models, upping displacement by five cubic inches per cylinder. The smallest of the set, the three-cylinder 2040, gained 15 cubic inches; the largest of the set, the six-cylinder 2940, gained 30 cubic inches after the engine block was strengthened. Power output remained about the same.

The Mannheim-built 2940 became the first utility tractor equipped with the new top shaft synchronized (TSS) transmission, providing 16 forward speeds and hi-lo shift. Buyers also had the option of equipping it with MFWD. German engineers had moved the front differential to the side and mounted the tie rod ahead of the front axle. The tractor was distributed in both the United States and Canada.

The 2040 and 2240 three-cylinder models, also built in Mannheim, could have an optional TSS transmission, a hi-lo direction-reverser, or MFWD.

Dubuque was building the two midsize 40 Series utility tractors. The updated 2440 and 2640 had an important 20 percent increase in lift capacity on the three-point hitch. The new 2640, introduced in 1980, could also be equipped with the hi-lo direction-reverser.

Schedule Masters

Canada's 1980 tractors in the John Deere 40 Series Schedule Masters line were designed to do more in a given amount of time. They were available with three-, four- or six-cylinder engines. They had some new engine features, a hi-lo option in all models, and redesigned wet disc brakes.

Continuing the Worldwide design approach, the Schedule Masters had a series of performance upgrades that were important for Europe as well as Canada. MFWD for the four larger utility tractors was the most important innovation, particularly for European sales.

Generally, MFWD boosted pulling power by 20 percent or more. It also improved stopping power, but reduced maneuverability and required a wider turning radius. Farming on steep hillsides was common in Europe. MFWD enabled braking with all four wheels. Previous utility tractors from Deere relied on a hydraulic-drive assist for steering, a feature which didn't permit mechanical braking.

The German-sourced Canadian models carried their own 40 Series model numbers, starting with the 44-horsepower 1040 and the 50-horsepower 1140. These Mannheim-built utility tractors were three-cylinder models ideal for orchards and vineyards and nearly identical to the 2040 and 2240 in the U.S. market.

Dubuque supplied four-cylinder engines for three Canadian models. They were designated as the 55-horsepower 1640, the 60-horsepower 1840 and the turbocharged 70-horsepower 2140. The Canadian 1840 was different from its U.S. counterpart, the 2440, in one key point: the 1840's muffler was hidden under the hood.

A sixth utility tractor for the Canadian market (nearly identical to the 2940 in the U.S. market) was the six-cylinder 80-horsepower 3140. In 1981, the SG2 Sound-Gard cab became available on the 2140 and 3140 tractors in Canada. It was built near Mannheim, at the new Bruchsal Works.

Overseas Tractors

Changes were happening in several countries in the late 1970s for Deere & Company. In Venezuela, a new tractor

Built in Mannheim and sold only in Canada, the 2140 was one of five tractors in the Schedule Master 40 Series. Major features for this tractor were mechanical front-wheel drive and a turbocharger that gave the tiny four-cylinder diesel about 70 horsepower. It was built between 1980 and 1987. This one was having a clutch replaced and a new seat installed at a Winnipeg dealership for a second generation of service.

factory opened in 1978. It became an assembly point for components made in Waterloo. However, the operation had to be closed two years later due to an economic downturn in the country.

In this period, John Deere also sold about a million dollars worth of equipment into a new market, China.

In Argentina, the factory at Rosario was closed in 1978 due to a farming depression. Dealers in Argentina began offering the smaller Mannheim-built 40 Series tractors (the 2140 through 3140). In 1982, the 4040 went into production and stayed for only a couple years.

In Australia, Deere dealers began selling the 1040 and 3140 tractors, as well as two models from Japan, the 850 and 950 tractors.

In South Africa, a late 1970s decree by the government ordered that diesel engines for tractors sold in South Africa be built in South Africa. Two years later, a new engine assembly facility was opened by a subsidiary of Perkins. John Deere brought in the Mannheim 40 Series tractor, installed the South African–built diesel engine, and marketed these as the 41 Series. The tractors ranged from a four-cylinder 1641 with 41 horsepower to a six-cylinder 66-horsepower 3141.

Farming Equipment

Company directors were convinced that construction of a dedicated R&D center for tractors in the 1960s had been a good investment. In the late 1970s, they committed to the industry's first dedicated center for combine technology. It was built in East Moline, close to the combine factory, and was opened in 1979.

Titan Combines

John Deere combine technology leaped forward in 1978 with the introduction of a new series in the United States and a second new series in Europe. Further upgrades to combine harvesting technology were offered in the U.S. market in 1982.

The new combines for North America were larger than previous models and were collectively known as the Titan series. They were described as the most productive lineup ever offered. Five models were offered, with variants for grain, corn and beans, and rice and soybeans.

The largest was a new model, the 8820 Titan, which produced up to 45 percent more output than any earlier

Bill Hewitt poses with this display model eight-row corn combine in the Deere & Company Administrative Center. Deere was selling more than 10,000 corn heads each year in the 1970s. They were so popular that a Nebraska company began making adapters so 40 series corn heads could fit on competitive combines. *Anna Wolfe collection*

John Deere combine. The 8820 Titan had the Waterloo-built 466-cubic-inch diesel engine, which produced 120 to 200 horsepower. The concave and beater grate area was increased 51 percent, and the straw walkers were 20 inches longer than those on previous combines.

Other models in the Titan series included the 6620, the SideHill 6620, the 7720 self-propelled, and the 7721 PTO combine. These offered up to 20 percent more capacity than previous models.

The updated 1982 Titan combines, with the aid of the new research center, had improvements such as a feeder-house reverser, Dial-A-Matic header height control, and a pressure gauge indicator for the accumulator.

Buyers had a choice of two new header platforms for the Titan combine. One was the rigid 222 for the

Introduced in 1979, the John Deere 7720 Titan self-propelled combine could be purchased with or without a hydrostatic rear-wheel drive for muddy conditions. This one is ready for the next harvest, wet or dry, at Dennis Kostenchuk's farm near Mountain Road, Manitoba.

The Titan II (or 8820) combine was released in 1985. Engine power could reach 225 horsepower. It had up to 15 percent more capacity and included a new feeder-house reverser. It has a 200 series, 30-foot platform that was easy to attach and offered automatic header-height control.

improved 6622 Hillside version. It was two feet wider than any previous platform for a hillside combine. The second, a new pickup platform, was 154 inches wide (nearly 13 feet). It was 22 inches wider than any previous John Deere pickup, and was ideal for the increasingly large windrows of grain in fields.

Cotton Pickers

America's cotton growers were startled by the 1980 introduction from John Deere of the first four-row cotton picker, the 9940. It became a powerful alternative to the two-row 114-horsepower 9910 introduced two years earlier and the less popular four-row 484 cotton stripper.

Field testing indicated the 207-horsepower 9940 could nearly double the productivity of a single machine in a field. It carried a high drum with a higher capacity, carrying up to 856 cubic feet of cotton. The shipping weight of the 9940 was 24,800 pounds, nearly double the weight of the earlier cotton picker.

European Combines

Deere had introduced a new top-of-the-line combine with six straw walkers for Europe in late 1978. The 985 was the first John Deere European combine with a 60-inch-wide separator.

In 1981, the entire European series of combines was replaced by the 10 series. Built at Zweibrucken, Germany, they had four, five, or six straw walkers.

Proven threshing features of earlier European combines were kept or improved. Engine horsepower was increased 15 percent on the Hydro-4 models. Grain tank sizes also increased about 15 percent. The new SG2 cab was adopted for these combines (as well as European tractors), along with new controls and instrumentation. A deluxe seat and air conditioning provided a new standard of harvesting comfort.

Tillage

The classic John Deere plow got a redesign and emerged in 1979 as a serious success. It was the equivalent of the New Generation tractor or New Generation combine.

The new family of plows offered a choice of new and more dependable safety-trips or dual-pivot spring-reset standards. They also had a new design. The bottoms were designed for plowing at four and five miles per hour. With the high-speed design came a reduced power requirement. These plows required 12 to 26 percent less draft than units from the two primary competitors.

Trash handling was improved on the 2600 and 3600, with higher underframe clearances (29.5 inches) and larger fore-and-aft clearance (27.5 inches). Cutting width was manually adjustable with shims for 16, 18, or 20 inches.

On the 2800, the clearance increased to 33.5 inches both fore and aft and under the frame. Cutting width could be adjusted hydraulically from 14 to 24 inches. Both the 2600 and 2800 had an exclusive hydraulic steering feature for the tail wheel.

Further refinements were offered in 1980 and 1982. The 2700 semi-integral plow, a hybrid, was offered in 1980. The flexible-frame 3700 drawn plow could have up to 16 bottoms.

Four units of this 9350 drill were for sale at Enns Brothers in Winnipeg. Built between 1975 and 1991, the 9350 was one of the first conservation drills from a full-line manufacturer. Disc openers produced minimum soil disturbance. It also had a high-capacity grain-and-fertilizer box for one-pass seeding.

Engineering teams quickly improved most types of farm machinery in the 1960s and 1970s. One example is this 14-foot 1380 mower-conditioner. It was tested and approved as very good in June 1980 by the Prairie Agricultural Machinery Institute located in Humboldt, Saskatchewan. It could operate at rates up to 7.4 acres per hour. Windrow formation and quality varied from good to very good. It also was easy to operate and adjust. This one is owned by Norman Friesen near Eden, Manitoba.

The plows were better than ever, but farming was moving in another direction in the early 1980s. Instead of burying all crop residue, growers were switching to chisel plows and discs for conservation tillage.

The new conservation farming tools were starting to save time, fuel, water, and soil. Moldboard plow sales in the United States dropped from 23,400 units in 1981 to only 9,100 the following year. At the same time, dealers sold 13,100 chisel plows in 1982. Today, most farmers no longer use moldboard plows.

Industrial Equipment

The 1970s were good for John Deere's industrial division. It had particularly strong growth in 1978, leading managers to think in terms of moving up in construction industry leadership as Deere had with farm machinery.

In 1978, Deere ranked fifth in sales of industrial and construction equipment. Caterpillar dominated the industry. Its worldwide sales were about nine times as much as Deere. Komatsu was second, with three times the sales volume of Deere. A German company, IBH Holding AG, was third. Even International Harvester was slightly ahead of Deere in sales.

On the other hand, Deere industrials were gaining ground. Earlier than Hewitt had predicted, in 1979, industrial sales edged up to 99.68 percent of $1 billion. The John Deere industrial division was building well over 2,000 pieces of industrial equipment each month, and supplying these to more than 500 dealer outlets worldwide. Sales were strongest in small and midsize equipment. The division sales in 1979 were more than a fifth of total Deere sales and generated profits of nearly $100 million. Feeling a little euphoric, management set out a goal to be "number three in the industry by 1990, and number two by 2000."

A market study in 1980 gave the company pause for thought. It revealed that Komatsu could market in the United States at very competitive prices while Deere was at a 25–33 percent cost disadvantage in the crucial Far Eastern markets that represented about 30 percent of industry sales.

An example of an ERA II John Deere industrial machine, this is the 450-B crawler at work in Aspen, Colorado, in 1972. *Deere & Company*

A year later, recession had hit the industry hard. By 1981, division sales had fallen to just over $780 million with a net loss of $38 million.

Despite having excellent manufacturing facilities for industrial equipment in Dubuque and a new factory at Davenport, Deere company managers decided after the recession to scale back the earlier goal. The industrial division decided to maintain its major positions as supplier to the forestry industry and in the small to medium-size markets for earthmoving equipment.

Forestry Machines

In 1978, the division had introduced several important machines that became strong performers. Machines for forestry were the most impressive and the largest. The JD593-B feller buncher and the JD743 tree harvester brought heavy industry into the forest for harvesting operations. In 1978, they were as much a leap forward as modern combines were to scythes for cutting wheat. Like the self-propelled combine, they are essential tools of today's forestry companies.

The feller buncher was based on the hydraulic excavator that carried an operator's cab over a set of dozer-type tracks and a long arm for digging. Instead of carrying a bucket for digging, the feller buncher had another tool for cutting trees. Tracks, in a forest setting, were more mobile than wheels. The feller buncher operator could swing the arm to a tree, clamp onto it, and shear it off at the base. Still clamping the tree, the operator could turn the arm and lay down the tree at a place for further processing. Seven track and two wheel models of the feller buncher are in production today.

The tree harvester on tracks also had an arm that could clamp onto and shear a tree. The machine carried a second tool, a spiked roller that drew the tree through a set of cylindrical knives and quickly stripped the branches. In less than a minute from start to finish, it propelled a stripped log to the rear of the machine. A forestry crawler with a grapple loaded the logs onto waiting trucks. Six track and five wheel models of the forestry harvester are in production today.

Construction Machines

The five-year ERA III program was completed in 1978 and 1979. New equipment for 1978 included the big JD890 excavator, the JD860-B scraper, JD670-A and JD770-A graders, and two large crawlers—the JD850 bulldozer and the JD855 crawler-loader.

Final machines in the ERA III cycle were the 1979 JD750 widetrack bulldozer and the JD844 articulated loader. The JD844 came with the company's biggest engine, a 955-cubic-inch V-8.

New factories, new electronic technologies, and new hydraulics enabled the John Deere industrial division to further upgrade its products in 1980. These new products were described as "a vision of things to come."

The JD862 scraper, released in April 1980, signaled the beginning of the company's high-tech contributions to construction. Long, hand-operated levers for controlling and guiding the machine were gone. In their place, the operator had finger-activated rocker switches at the end of padded armrests. Out of sight, a microprocessor controlled six sensors. This would determine the best shift sequence, shift the transmission automatically, control the torque converter, protect the whole mechanism, and diagnose problems. The goals were maximum efficiency and operator comfort.

For the 1980 Olympic Games in Lake Placid, New York, John Deere was the equipment supplier. These are just two pieces of the fleet. *Deere & Company*

Deere needed more exhibit space than ever at the early 1981 CONEXPO show in Houston. Machines now were identified without the "JD" in front of the numbers. The name John Deere was placed in larger, attractive decals separate from the model number.

The new models at CONEXPO included the 762-A scraper, the 655 and 755-A hydrostatic crawler-loaders, a model 646-C compactor, and four four-wheel-drive loaders—the 444-C, 544-C, 644-C, and 646-C. The full line of industrial equipment from John Deere now included 68 models. There were 17 forestry machines, 22 construction machines, and 29 utility vehicles, as well as various attachments and associated equipment.

Along with the equipment, in February 1981, the John Deere Training Center was opened in a new 55,000-square-foot facility near the Davenport Works. The center provided backup support for industrial and agricultural equipment worldwide. The company already had field demonstration and training facilities for equipment in Illinois, Georgia, and Arizona.

The Dubuque and Davenport factories had 168 acres under roof in 1981. Additional John Deere industrial equipment was being built at factories in Germany, France, and Australia. The company had more than 650 dealer outlets, including 234 overseas. Deere & Company served about 80 percent of the world market.

Not quite new, this John Deere 90 skid steer is one of the early units. It came out of the factory between 1978 and 1984 and now works most weekdays for Canotech Consultants Limited of Winnipeg, Manitoba.

Deere market share for fiscal year 1981 was 20 percent for industrial wheel tractors and four-wheel-drive loaders; 28 percent for wheel log skidders; approximately 30 percent for crawler dozers, crawler loaders, and motor graders; and 36 percent for elevating scrapers under 18 cubic yards.

Consumer Products

Deere & Company had established a consumer products organization in 1963, starting with a single 110 lawn and garden tractor introduction, built at the Horicon Works factory in Wisconsin. Consumer products eventually became the fastest growing business segment for the company.

By 1978, John Deere machinery dealers were seeing sales generated year-round as customers selected from 88 separate pieces of equipment for grounds care, landscaping, gardening, or weekend farming. In the next decade, Deere became the world's largest maker of lawn and garden tractors and soon offered the most complete line of new grounds-care products available anywhere for residential and commercial users.

Grounds-care equipment at the local John Deere dealer included utility tractors and loaders, backhoes, snow blowers, skid steer loaders, and much more.

Yanmar Tractors

Japan became a supplier of new compact tractors late in 1977. Japanese manufacturer Yanmar began distributing John Deere tractors in Japan in 1976. In turn, Yanmar supplied Deere with its first compact diesel tractors, filling a power gap in Deere's North American line of equipment.

In 1978, the 22-horsepower 850 and 27-horsepower 950 tractors, based on tractors well proven in Japan, were made available through Deere dealers in North America. These stretched the line, giving dealers a stepped selection of utility tractors from 22 PTO horsepower at the low end up to the model 2240 with 50 PTO horsepower. A third diesel compact was added late in 1979, the 33-horsepower 1050.

The Yanmar tractors had three-cylinder diesel engines. The modified North American John Deere versions had more cooling capacity and could travel faster than matching models used in Japan's rice fields. They also had a clutch to override the 540-rpm PTO and a two-post Roll-Gard structure. The operator's station was resized to match the size of customers in North America, and Generation II styling was added. The tractors had 30 pieces of matched equipment, but most were sold with a mowing deck.

Horicon Tractors

Below the utility tractor line, John Deere dealers could offer a line of lawn and garden tractors from the Horicon Works. The smallest was a gear-drive 108 or 111 lawn tractor. The 111 was joined in 1981 by a hydrostatic version for small acreage users. One lever controlled its direction and speed.

Slightly larger needs could be met with a 16-horsepower 116 lawn tractor equipped with either a 38- or 46-inch mower deck. Options improved the next year. Buyers then could select an 18-horsepower 318 or a

In 1978, a Japanese manufacturer, Yanmar, supplied the first John Deere compact diesel tractors—the 850, 950, and 1050. This John Deere 950 is owned by the Odanah Hutterite Colony of Minnedosa, Manitoba. It's a great piece of equipment for a garden that feeds a hundred people.

20-horsepower 420 lawn and garden tractor powered by the Onan two-cylinder air-cooled engine, or they could go to the Japanese-built "Taskmaster" line. This line included two- or four-wheel-drive models with 14.5 PTO horsepower up to the 40-horsepower three-cylinder 1250 offered in 1982.

Total sales in the consumer products division from 1979 through 1981 amounted to a third of a billion dollars each year, and the products remained profitable. Profit from consumer products was running at 11 percent in the 1970s. In 1980 and 1981, it dropped to less than 8 percent, but still bolstered the bottom line for Deere & Company during the recession.

The strongest growth in the division was in commercial mowing products; the least growth was in snowmobiles. When the economy began to recover, Deere still had about 25 percent of the market for lawn and garden products.

The Yanmar 27-horsepower three-cylinder engine had a 105-cubic-inch displacement. A full line of field equipment could be operated by the tractor. It was available as a two-wheel-drive or a four-wheel-drive with MFWD. The tractor weighed just 2,700 pounds and had a wheelbase of only 69 inches. It continued in production into 1988.

The 146-horsepower six-cylinder John Deere 7020 was manufactured at Waterloo, Iowa for model years 1971 through 1975.

PASSING THE TORCH

Power Shifts

The year 1960 marked the start of the New Generation tractors and the passing of the two-cylinder tractor era. In a similar vein, 1982 marked the end of one era and the start of another: Bill Hewitt's leadership at the helm of Deere & Company ended and the leadership of Robert Hanson began. Hewitt had been presented with an old, family-owned, well-managed machinery company that had both feet firmly rooted in Midwestern soil. The company he passed on to Hanson was among the best-managed multinational and multifaceted corporations in the world, still with its roots in the Midwest.

The three sixth-generation descendants of John Deere had flown the nest. Adrienne Deere Hewitt graduated Phi Beta Kappa from Dartmouth College in 1977. Adrienne was attending Georgetown University law school in 1980 when her twin sister Anna married a Moline native, Joseph Vincent Wolfe. In 1981 and 1982, their younger brother Sandy earned bachelor's degrees at Knox College in economics and physics. As his father retired from Deere & Company, Sandy went to work for Chase Manhattan in New York City.

Robert A. Hanson had begun working at John Deere Intercontinental in 1950, about the same time that Hewitt moved to Moline as a newly elected company director.

Hanson was born in 1924 and raised in Moline. In 1973, he was elected senior vice president of Deere's Overseas Division. He caught Hewitt's attention and became one of four executives considered by Hewitt and Elwood Curtis as candidates to become the next company president. Two years later, Hewitt announced his choice of Hanson for the newly reinstated post of executive vice president. The motion was accepted, and Hanson became the designated successor.

Promoted to president of Deere & Company in 1978, Hanson became the first company president who wasn't related by blood or marriage to the founder of the company, John Deere. William A. Hewitt retired in 1982, accepting a post as United States Ambassador to Jamaica.

John Deere's colorful magazine, *The Furrow*, was published worldwide with many editions in many languages in the 1980s.

Hanson was then elected chairman and chief executive officer in October 1978. Hanson held the helm for eight years and retired in 1990.

Faltering Competitors

Coming out of the 1970s, Deere & Company was in the strongest market position in the agricultural machinery industry. Profitable divisions of the company also built earthmoving equipment and outdoor consumer products. Companies that had led or rivaled Deere in the farm machinery industry two decades earlier were in trouble. In 1981, Deere's market share of United States row-crop tractors was more than 45 percent; in Canada it had almost 31 percent market share. Deere & Company was honored as one of the best-managed firms in North America in 1981, and this was the year the industry changed.

Deere managers had focused on good employee relations and internal efficiency since the 1930s, when it was one-eighth the size of International Harvester. The only extended strike action had occurred in late 1967 as the United Auto Workers (UAW) union insisted on having a single master agreement covering all the Deere factories in the United States. That strike began in late fall and was settled in early January of the following year. For most of the next five decades, gross profits at Deere were about 70 percent higher than at IH; some incremental profit rates were roughly double what IH achieved. International had been the clear industry leader when Hewitt set his goal on being number one, but the company was in trouble before the 1980s recession.

International Harvester

International Harvester had never achieved Deere's focus on efficiency and production. In the postwar era, International Harvester tried to be all things to all people. The company built trucks, construction equipment, and even home appliances while continuing with international sales and setting new goals. It was having serious labor troubles, and was being more generous than it could afford with employee packages. After 1966, when new management tried to turn it around, it was too late.

International Harvester was building these powerful 2+2 tractors from 1981 to 1984, as it was going through a disastrous economic collapse. The company built only 605 of the 177-horsepower 6588 tractors. Because of the long hood, the machines also became known as the "Anteater" or "Snoopy" tractor. Ed Wiebe, a former IH dealer, has kept this example near Carberry, Manitoba.

The pivotal event came in 1979. The UAW union had contracts up for renewal with IH, Deere, and Caterpillar. All three companies experienced UAW strikes. At Deere, the strike was settled in 21 days. The settlement gave a modest wage increase but tightened up provisions for the sickness benefit program. At Caterpillar, the strike lasted about 80 days. It was acrimonious and complicated by new leadership at the UAW local branch.

At International Harvester, the strike was disastrous. It shut down the company for 172 days. Harvester's new CEO had come into the office just a year earlier with a reputation for being fiscally oriented. The union had a $90 million strike fund on hand, and when the CEO pushed for cost-cutting changes in the new contract, both sides took a hard stand. By the time it was settled and employees were back at work in the mid-1980s, the U.S. was in a recession that reduced farm equipment sales and decimated heavy-duty truck sales. At the end of 1980, some 200 banks and insurance companies put up a $4 billion rescue plan for International Harvester, but it wasn't enough.

International Harvester lost $2.4 billion between 1979 and 1982. The new CEO resigned in 1982 and the directors sold the construction equipment division.

In 1985, the company's agricultural equipment business, along with the "International Harvester" name, was sold to Tenneco, a highly diversified investment company.

J. I. Case

A second competitor in the agricultural machinery business, J. I. Case, owned roughly 5 percent of the tractor market in the 1950s. It was a pioneer in threshing equipment, but Case entered the 1960s in bad shape. Ownership changed in 1964 and again in 1967. That year Tenneco acquired 53 percent ownership of J. I. Case, along with other assets of the previous owner. By 1970, with about 4 percent of the market, Case ranked sixth in the North American agricultural machinery industry. It withdrew from making farm implements in 1970 and, in 1972, discontinued its line of combines. At that point, Case was no longer a long-line farm equipment manufacturer, although it continued building tractors and enjoying good sales in construction machinery.

Tenneco consolidated its ownership of J. I. Case in 1976 and supported Case operations during the recession. In 1985, after the recession, Tenneco invested roughly $430 million to purchase the remaining International Harvester agricultural assets. Tenneco combined the International Harvester operations with its Case subsidiary and created Case IH. Tenneco officials believed Case could benefit from Harvester's broad product line and stronger dealer network. The combined group then commanded a 35 percent market share for large tractors—second only to Deere & Company—and became a new competitor to Deere.

Massey-Ferguson

A third competitor, Massey-Ferguson, had collapsed into near bankruptcy in the late 1970s. It once was the most powerful agricultural machinery builder in Canada, with subsidiary companies around the planet, but it sank slowly under a sea of problems.

The Massey family lost control during the 1920s when it floated the company on the Toronto Stock Exchange. After World War II, it was managed by professional boards and committees, with the singular goal of maximizing earnings without regard to long-range strategic needs. The company borrowed heavily for investment

The J. I. Case Company built this sturdy and stubby articulated four-wheel-drive from 1979 to 1983. The Case 4690 had a class III hitch and was rated at close to 220 horsepower. It had a 102-inch wheelbase, about two feet shorter than the Deere 8440, and turned a tighter circle.

Massey-Ferguson built the 320-horsepower (PTO) 4900 articulated four-wheel-drive tractor from 1980 to 1983. The model was powered by a Cummins eight-cylinder diesel and had a base price of $123,000. Deere's largest model at the time was the 8640, with six cylinders, 228 PTO horsepower, and a retail price of $80,000.

Ford's solution to re-entering the North American articulated four-wheel-drive tractor market came in 1987 when it purchased Versatile Manufacturing Limited, based in Winnipeg. The Versatile name remained on this 846, under the 1989 Ford moniker, and the tractor was painted Ford blue.

capital, and then encountered a combination of record high interest rates and an economic slump.

In 1978, before the recession, it lost $262 million. A massive financial bailout, involving 250 banks and insurance companies worldwide, plus governments in Canada and Great Britain, wasn't enough to save the company. New investors began a long reorganization process. The remaining assets reached firm ground again in 1995, as a subsidiary, when Massey-Ferguson's worldwide holdings were purchased by the U.S.-based AGCO Corporation.

Other Competitors

A fourth competitor, Ford, had about 19 percent of the tractor market share in the 1950s, putting it somewhat ahead of Deere & Company. Ford introduced a much more convenient power shift transmission, the "Select-O-Speed," soon after the New Generation was introduced. Select-O-Speed also featured an advanced hydraulic system to compete with the Deere closed-center system.

Challenged by Select-O-Speed, Hewitt sent his Product Development Center staff into a spare-no-expense crash program to develop a better shifting system. It was introduced as "Power-Shift" on the flagship 3010 and 4010 tractors of 1963. Deere Power-Shift surpassed

customer expectations, while Ford Select-O-Speed had issues due to insufficient development, and soon was the less-favored system.

In 1964, Ford pursued the lucrative world market for high-volume utility tractors. The marketing, to this point, had been dominated by Deere and International Harvester. Ford first merged its separate British and American tractor lines in 1965. This led to a dynamic period with new models and sophisticated features.

Ford then launched its own worldwide tractor line in 1965. The following year, Ford worldwide sales nudged ahead of International Harvester and placed Ford as the world's second-largest farm equipment manufacturer. Ford survived the recession, helped by sales of a new four-wheel-drive model built by Steiger Tractor at a plant near Fargo, North Dakota. In 1985, Ford production of four-cylinder and TW Series tractors was transferred to Europe.

Allis-Chalmers had about 10 percent of the tractor market in the 1950s, but tractors were only a very small part of this very large company in the 1960s. All Allis-Chalmers tractors, with one exception, were produced at the factory in West Allis, Wisconsin. In that factory, farm tractors only took about a quarter of the floor space and were built on dated assembly lines. One observer recalled that despite heroic efforts to infuse capital and to modernize, the old-time line-shaft support structures hung from the ceiling to the end of the company.

After New Generation tractors changed the industry, it took 13 years for Allis-Chalmers to develop and release a competitive tractor. Meanwhile, the industrial giant was building machinery for cement, mining, and hydroelectric projects, flour mills, and sawmills, as well as products for refrigeration, home heating, water softeners, and nuclear power.

The 6000 and 8000 diesel tractors introduced during the 1981 farming recession were the last pure Allis-Chalmers tractors. They were a good product without the dealer organization to support high-volume sales. Most of the tractors sold went to small, low-volume mom-and-pop outfits.

In 1985, Allis-Chalmers was sold to Klockner-Humboldt-Deutz AG (KHD) of Germany. KHD planned to sell its German-built Deutz tractor through Allis-Chalmers dealerships, but soon discovered the main reason for Allis' stagnation. Then KHD learned that its European tractor, tuned to European needs, wasn't fully suited to the North American market. That became the final nail in the coffin for Allis-Chalmers.

Three other well-established tractor-makers were competing in the 1950s—Oliver, Minneapolis-Moline, and Cockshutt. Together, they had about 10 percent of the market. Gradually, over the next two decades, the individual company names and colors disappeared as they became part of the White Motor Company.

After 1975, White stopped using the other names and painted all the White tractors in a two-tone gray. Then the White Farm Equipment Company was hit by hard times. It went bankrupt, and its farm equipment division was sold in late 1980 to TIC, a Texas investment company based in Dallas. Stability was regained after the 1987 formation of White-New Idea Farm Equipment Company.

Factory of the Future

During the prosperous 1970s, Deere & Company invested heavily in technology in a way that broke new ground for heavy industry in North America. Deere's "factory of the future" concept brought together the best technology in the world into a purpose-built production facility. The purpose: to build farm tractors.

The assembly-line tractors of previous decades had been mostly built by hand in old factories. Gradually, tractors became complicated. More parts were required. Extensions and wings were added, assembly lines were modified, and tools were conceived and built. To the outside observer, the 1960s tractor factory was a miracle of orchestration. Out of the cacophony of activity, noises, and smells, gleaming new tractors somehow emerged.

By 1970, Deere & Company's engineering managers could see that more capacity would be needed than they could build into or provide at the old Waterloo Works

The Allis-Chalmers company built these 4x4 articulated tractors from 1982 to 1985. The 4W-305 produced 254 PTO horsepower. This rare, well-kept model is owned by David Minkus in Minnedosa, Manitoba, and was on exhibit at the 2009 Neepawa Lily Festival.

facility that had served the company for more than 50 years.

About that time, Deere developed a daring and financially risky idea, according to historian Wayne Broehl Jr. The idea was to custom-build the factory of the future with amazing machines and more production efficiency than ever seen before. They would need more than one building, and they estimated it could cost $200 million a year and take about ten years to build. Hewitt's response was, "We've got to do it—get started."

Engine Works

The first step was a new engine factory. At the conclusion of a four-day retreat back in southern California in 1974, the board agreed to continue manufacturing engines at Waterloo while the new facility was being built. They purchased 139 acres southwest of Waterloo and started construction. The first engine came off the new assembly line in 1976.

John Deere Waterloo Engine Works was the most modern engine factory in the world. According to a 1999 description, the factory and office complex had 21

White Machinery dealers sold the 240-horsepower (PTO) 4-270 from 1983 to 1988. The tractor was powered by a six-cylinder Caterpillar engine. This model was purchased new in 1987 by the current owner, David Adams of Bagot, Manitoba.

acres under roof and was a third of a mile long. The factory could build 100 different engines simultaneously, and functioned with a staff of about 850 workers plus a handful of robots that handled tasks like spray painting and loading parts in heat-treating operations. It was supported by a new nearby electric foundry that had begun operating in 1972. The foundry's annual capacity was increased in 1975 to 236,000 tons.

Product Engineering Center

An updated Product Engineering Center (PEC) was commissioned in Waterloo in 1977, at about the time that the new Iron Horses were introduced. Employing 125 people with advanced professional and technical degrees, the PEC had the air of a university campus in a cornfield.

The PEC technicians and assistants could test engines, transmissions, hydraulic systems, and electronic components on site after the initial design. There were cold rooms, hot rooms, sound rooms, and wind tunnels. Scientists could take a tractor and blow on it, broil it, bump it, freeze it, pound it, shake it, and twist it for the purpose of producing the highest-quality product.

The lab also enabled Deere to pay attention to operator safety, noise levels, pollution emissions, and other human issues.

Deere's large-capacity testing facility at Waterloo set the stage for further expansion of Deere's industry leadership. The PEC lab became the technical laboratory of choice for other heavy-equipment manufacturers and the source of frequent proposals for internationally recognized technical standards.

The PEC became a leading lab in the U.S. for the use of computer-assisted design and engineering. Computers could generate design specifications for feeding to machine tools to build new components. Software was developed so that it was possible to search an inventory of 250,000 parts, looking for existing similar parts while a new design was in progress.

In the mid-1990s, Deere acquired Phoenix International of Fargo, North Dakota, to extend its engineering research capability. Phoenix is an advanced developer and manufacturer of microprocessor-based electronic control equipment for several manufacturers worldwide. Deere continues to be at the forefront of new technology for farm and off-highway equipment.

Tractor Works

The most difficult building, and largest, was a tractor assembly facility. In 1975, the company purchased a large corn farm on the northeast edge of Waterloo. Construction began in 1977 on the 1,300-acre site. Elements of tractor assembly began at the new, ultra-modern, single-story factory in 1979, but it would take another two years to complete the project. The site became fully operational in May 1981.

The finished tractor assembly complex occupied 2,100,000 square feet (48.2 acres) with seven buildings under one roof. It incorporated the latest concepts in material flow, process sequencing, and process routing. Overhead, a 7.5-mile computerized crane transportation system handled parts delivery. The land surrounding the buildings provided parking and space for testing equipment. A large portion of the former farm was seeded with native prairie grasses.

After the factory of the future was completed, an outside management expert observed and described the new system. It had a systemic impact on production, reaching from factory worker to dealer to corporate manager.

Superficially, a computer directed hundreds of "robocarrier" carts to the parts bins—about 10,000 stacked wire cages. Each robocarrier could carry a 4,000-pound payload at up to 135 feet per second. If a worker got in its path, it would stop until the path was clear again. The system enabled Deere to reduce inventory by about half.

The system also enabled the plant to assemble customized tractors. Somebody figured out that Waterloo Tractor Works was able to turn out 5,000 different tractor configurations without interrupting production.

The new assembly facility permitted Deere to reorganize the old downtown factories in Waterloo. They became facilities for making major components, with different crews assigned to building transmissions, hydraulics, or gears. Workers were subdivided into work crews assigned to a given component. They could feel they were part of a small company within the larger complex. For instance, a crew of three would run a cluster of lathes, hobs, shapers, a shaver, and a milling machine to make a single gear. This system streamlined the production of that component.

Deere installed numerically controlled machines that were linked into a computer network. The computer could reprogram a machine to do a different task; it also could rationalize the entire system by directing parts and work to each machine at an ideal pace. This reduced machine idle time and parts queues.

During the process of modernizing production and assembly, the engineering group came up with a system to better use the parts that were produced. Called "group technology," it became a system for identifying parts based on shapes and function into many families of parts. This enabled Deere to standardize and reduce the overall number of parts in inventory. It also reduced the cost of handling, allowing a family of parts to be manufactured by common tools. The result: in the past, only 7 percent of the parts used in one manufacturing department were made completely within that department; afterward, using group technology, 72 percent of the parts used in one department were made completely within the department.

Overall, Deere stressed teamwork among workers using a concept of a factory within a factory. There were no layoffs as a result of the technology. When the recession hit there were layoffs, but the new factories remained able to break even while operating at only 40 percent of capacity. In 1980, Deere weathered the worst farm recession since the 1930s without experiencing a losing year. During that time, it boosted both capacity and efficiency and was poised to move ahead in market share when the economy recovered.

According to author James O'Toole, Hewitt believed that the future belongs to those who think systematically. O'Toole records this statement by Hewitt: "We have no special advantage that cannot accrue to any other company. The primary way we can maintain and advance our position is through better planning, design, engineering, fabricating, distributing, selling, and servicing—through better work on the part of John Deere people and groups."

The West Office Building addition to the Administrative Center, designed by Eero Saarinen's successors, was cited as a second architectural masterpiece. *Kevin Roche John Dinkeloo and Associates*

John Deere People

As we close the story of Deere & Company in the 1960s and 1970s, a few last threads need to be picked up.

Throughout the decades from 1900 onward, Deere maintained good to exceptional employee relations. The UAW found it necessary to strike at least twice during Hewitt's tenure (1967 and 1978), but the relationship was mostly amicable. Deere enjoyed a strong level of worker loyalty; people wanted to work at Deere & Company, and stayed on because of the good working relationships. For instance, as an incentive in 1977, Hewitt approved the offer of a stock purchase plan for office workers. For shareholders, Deere stock provided a reliable modest return on investment.

Reflecting the rise of Deere & Company to a worldwide leadership position in farm machinery, the little company magazine initiated in 1895 for Midwest customers continued to grow and to reflect the changing company.

During Hewitt's later leadership, directors supported *The Furrow* with a large, award-winning, professional editorial staff and extended the magazine's international service. *The Furrow* expanded in the late 1970s to include a full-time editor for international editions and to include editions for Asia, Australia, southern Africa, Denmark, Norway, and Sweden. Staff produced three editions for Canada as well as many regional editions in the United States, including a new *Dairyland* edition.

Each edition was composed of some 200 pages of material supported by brilliant full-color photography. From that total collection, pages and ads were selected to comprise the regional editions. Today the magazine is published in 14 languages and distributed in more than 115 countries, making it one of the largest farm publications in the world.

There's a legend that John Deere provided credit to the purchaser of the third plow he ever sold. The fact is, the John Deere Credit Company was established during Hewitt's early tenure in 1958 to provide funding for retail financing programs. It was soon financing a large portion of Deere's farm equipment business. Ten years later, Deere acquired an insurance company in New York and formed the John Deere Insurance Group.

In 1978, the credit service was extended with an affiliate office for Canadians when Deere & Company opened a Canadian headquarters at Grimsby, Ontario.

Company credit and insurance were particularly important for dealers and customers when the farm economy took a nosedive in 1980 and 1981. In 1981, the two services generated a modest profit while spreading risk. Retail credit earned $57.7 million and insurance earned $27.2 million for the equipment manufacturer.

In 1978, in addition to the new John Deere Engine Works at Waterloo and the new Canadian headquarters, new buildings were dedicated in Moline and Atlanta. In Atlanta, Deere & Company opened sales branch offices, and in Moline, Deere's own "West Wing" was dedicated. Officially, it was the West Office Building addition to the Administrative Center, and was designed by Saarinen's successors, Kevin Roche and John Dinkeloo. It became another example of company excellence and quality. Deere West was visually stunning while enhancing the working environment. It connected to Saarinen's

masterpiece with a 194-foot fourth-level bridge. It featured an interior atrium with offices on all sides looking out over adjoining gardens. Like its predecessor 15 years earlier, it was cited as "Office of the Year" by *Administrative Management Magazine*.

In 1979, employment at Deere & Company worldwide reached an all-time high of 65,392. Sales topped $5 billion and earnings reached a record $310 million.

Corporate Review

The last four years of Chairman Bill Hewitt's 27-year leadership saw the continuing release of new technology from the divisions of Deere & Company while it weathered an economic typhoon. Former full-line farm equipment manufacturers floundered and their assets were reformed into new machinery companies. Deere & Company retained its identity. As the storm eased,

Ambassador Hewitt visited Cape Town, South Africa, in September 1995. After meeting Nelson Mandela, he noted, "It's amazing how dynamic Mandela was after having spent 27 years in prison. He was physically fit and in high spirits, with a sense of humor as well. In my book, he is a modern-day hero." *Anna Wolfe collection*

The American ambassador to Jamaica, William Alexander Hewitt, bidding farewell to Queen Elizabeth and Prince Philip in 1983. *Anna Wolfe collection*

Model 4020 often is regarded as one of the most outstanding tractors of the past century, and it is closely tied to the success of John Deere in the 1960s and 1970s. The restored John Deere 4020 row-crop tractors pictured here feature Roll-Gard structures. *Tony Gerber*

it was the only surviving, long-line farm equipment manufacturer that had not been acquired by an investment company or other multinational group or merged with a competitor.

Investors could only be pleased.

Deere & Company corporate life had begun in 1868. The corporation first issued 400,000 shares of preferred stock in 1911, raising capital to "produce a tractor plough." In 1955, when Bill Hewitt became president, the family controlled the board of directors and shares of common stock could be purchased for about $34.50.

By 1973, the company was widely owned by institutions and the general public. The descendants of John Deere still owned just over 12 percent of company shares and practiced block voting. Virtually everything in their world had changed in the Baby Boomer era, and the directors decided it was time for Deere & Company leadership to change.

The directors appointed the first outside director to the board that year. The next year, the unifying element of a family vote was surrendered as Deere family shares were dispersed into individual hands for voting purposes. The course was settled in 1975 when outsider Robert Hanson was designated as Hewitt's successor and four more outside directors were appointed to the 15-member board.

In January 1981, with the recession easing and a huge investment in new manufacturing assets, the company issued four million shares of common stock. Purchasers paid more than $41 per share. By the end of 1981, through stock dividends and share splitting, 100 shares of stock from 1955 would have multiplied to 824 shares with a market value of $25,995. Twenty-six years of dividend payments would have provided $14,771 to a shareholder. By 2009, there were more than 420 million outstanding shares of Deere common stock.

A portion of Bill Hewitt's success is attributed to his personal level of control and the continuity of his leadership through nearly three decades. The combination, unusual in a large corporation, allowed him to pursue goals often not allowed in the business world. Within the corporation, he trained the management team that would continue after his retirement. The strong leadership has carried on successfully long after Hewitt's personal stewardship.

50 Series Tractors at New Orleans, 1982

The number of tractors on U.S. farms had peaked in 1965 and by 1980 had fallen to 4,324,000, but total horsepower continued to rise. The estimated 248-million-horsepower total on farms in 1980 was higher than ever. Deere provided seven additional tractor sizes above 100 horsepower in the 1970s, and by 1982 had thirteen 40 Series models with 40 horsepower (PTO) and above.

The new 50 Series, introduced in July 1982 in New Orleans, linked innovation and new field demands. It rivaled the historic "D-Day in Dallas," twenty-two years earlier, when the first New Generation three- and four-cylinder tractors were introduced. Dealers were flown in from all over the world to witness the unveiling of ten new row-crop tractors in the spectacular setting of the Superdome.

The 50 Series had more horsepower (from five to ten horsepower, depending on the model) and all were available in mechanical front-wheel-drive versions.

Five tractors in the 100–190 horsepower range had a revolutionary new 15-speed Power-Shift transmission. It gave the operator extremely efficient operation, with an estimated 7 percent increase in fuel economy. Seven forward speeds were in the field working range (with choices of speeds from 3 to 7.5 mph), four were in the PTO range (below 3 mph), and four were in the faster transport range.

A second feature of these five models was Caster/Action, an innovation for the mechanical front-wheel-drive (MFWD) axle. One of the problems in MFWD adaptations for row-crop tractors had been its increase in the turning radius. With the Caster/Action principle, the front wheels cant as they turn, dipping and giving the effect of turning under a bit, just enough to allow a substantially shortened turning radius. Mechanically, the configuration also lends itself to integration with the drop-style, multi-speed Power-Shift transmissions produced long term by Funk, a Kansas subsidiary of Deere & Company.

Hewitt, spouse of a fifth-generation Deere family member, steered the old company over a rapid transition in the 1960s and 1970s. Deere became the largest agricultural machinery producer in the world as well as a major force in construction equipment and outdoor consumer products. Company management went worldwide, the workforce increased, factories were built and rebuilt, and a strong reputation for quality in product, style, and management was formed.

The factory of the future represented one of the most important achievements of the past 26 years for Hewitt and Deere. His final major event with Deere & Company came in July 1982 in New Orleans, as he unveiled a new line of advanced 50 Series tractors.

Hewitt was appointed U.S. Ambassador to Jamaica on September 30, 1982, by President Ronald Reagan and served until October 14, 1985. Other presidential appointments for Hewitt included the Special Committee on United States Trade Relations with East European Countries and the Soviet Union (1965), the National Advisory Commission on Food and Fiber (1965–1967), the National Corporation for Housing Partnerships (1968), the Task Force on International Development (1969), the National Council on the Humanities (1975–1980), and the President's Commission for a National Agenda for the Eighties (1980–1981).

Hewitt also served as the vice-chairman and chairman of the National Council for the United States–China Trade Commission, a board member of the California Institute of Technology, a member of the Sanford Research Institute, and a director for American Telephone and Telegraph, Continental Bank of Chicago, and Continental Oil.

BIBLIOGRAPHY

Baumheckel, Ralph, and Kent Borghoff. *International Harvester Farm Equipment Product History 1831–1985*. St. Joseph, MI: American Society of Agricultural and Biological Engineers, 1997.

Broehl Jr., Wayne G. *John Deere's Company: A History of Deere & Company and Its Times*. New York: Doubleday, 1984.

Dean, Terry, and Larry Swanson. *Antique American Tractor and Crawler Guide*. Osceola, WI: MBI Publishing, 2001.

Dietz, John. *Classic John Deere Two-Cylinder Tractors: History, Models, Variations, & Specifications 1918–1960*. Minneapolis: MBI Publishing, 2008.

———. *John Deere Two-Cylinder Tractor Buyer's Guide*. St. Paul: MBI Publishing, 2006.

Fay, Guy. *International Harvester Tractor Data Book*. Osceola, WI: MBI Publishing, 1997.

Gay, Larry. *A Guide to Hart-Parr, Oliver and White Farm Tractors 1901–1996*. St. Joseph, MI: American Society of Agricultural and Biological Engineers, 1997.

Henshaw, Peter. *Tractors*. St. Paul: MBI Publishing, 2002.

———. *The Ultimate Guide To Tractors*. New York: Book Sales, 2003.

Hobbs, J. R. *The John Deere 10 Series, New Generation Tractors*. Bee, NE: Green Magazine, 2004.

———. *The John Deere 30 Series, Second Edition*. Bee, NE: Green Magazine, 1994.

Klancher, Lee. *Farmall: The Golden Age 1924–1954*. St. Paul: MBI Publishing, 2002.

Leffingwell, Randy. *America's Classic Farm Tractors*. Osceola, WI: MBI Publishing, 1999.

———. *John Deere: A History of the Tractor*. St. Paul: MBI Publishing, 2006.

MacEwan, Grant. *Power for Prairie Plows*. Saskatoon: Prairie Books, 1971.

Macmillan, Don, ed. *The John Deere Tractor Legacy*. Stillwater, MN: Voyageur Press, 2003.

———. *John Deere Tractors and Equipment Volume 2, 1960–1990*. St. Joseph, MI: American Society of Agricultural and Biological Engineers, 1991.

Magee, David. *The John Deere Way: Performance that Endures*. Hoboken, NJ: John Wiley & Sons, 2005.

Marsh, Barbara. *A Corporate Tragedy : The Agony of International Harvester Company*. New York: Doubleday, 1985.

Meyer, Faith Hamilton. *John Deere Dubuque Works, 1947–1997: Changing Perspectives*. Dubuque, IA: John Deere Dubuque Works, 1977.

Miller, Merle. *Designing the New Generation John Deere Tractors*. St. Joseph, MI: American Society of Agricultural and Biological Engineers, 1999.

Nelson, Selmer and Nadine. *Grossenburg's Fifty Years with John Deere 1937–1987*. Sioux Falls: Pine Hill Press, 1987.

O'Toole, James. *Vanguard Management: Redesigning the Corporate Future*. New York: Doubleday, 1985.

Peterson Jr., Chester. *American Farm Tractors in the 1960s*. St. Paul: MBI Publishing, 2004.

———, and Rod Beemer. *Inside John Deere: A Factory History*. Osceola, WI: MBI Publishing, 1999.

———. *John Deere New Generation Tractors*. Osceola, WI: MBI Publishing, 1998.

Pripps, Robert N., and Andrew Morland. *Great American Tractors: Deere, Farmall, Ford*. St. Paul: MBI Publishing, 2003.

Rukes, Brian. *American Farm Tractor Dealerships*. Iola, WI: Krause Publications, 2004.

Sanders, Ralph. *Vintage Farm Tractors: The Ultimate Tribute to Classic Tractors*. Stillwater, MN: Voyageur Press, 1996.

Swinford, Norm. *Allis-Chalmers Farm Equipment 1914–1985*. St. Joseph, MI: American Society of Agricultural and Biological Engineers, 1994.

———. *The Proud Heritage of AGCO Tractors*. St. Joseph, MI: American Society of Agricultural and Biological Engineers, 1999.

———. *A Century of Ford and New Holland Farm Equipment*. St. Joseph, MI: American Society of Agricultural and Biological Engineers, 2000.

Wendel, C. H. *Great Farm Tractors*. Winnipeg: Bison Books, 1995.

Will III, Oscar H. *Cub Cadet: The First 45 Years*. Bee, NE: Hain Publishing, 2005.

Two-Cylinder magazine, V 10.1; V 16.4; V 17.2, 5, 6; V 18.2, 3, 4; V 20.4; V 21.1. Grundy Center, IA: Two-Cylinder Limited.

ACKNOWLEDGMENTS

Many thanks to the authors who have gone before me, who did the hard slogging through archived files and the long interviews with old-timers who were on site when events transpired, and who previously assembled these precious bits into their own works. Among them, Wayne G. Broehl Jr. compiled and wrote an 870-page history, *John Deere's Company*, which was published in 1984. The last chapters of that book covered the era portrayed in this work in significant detail, and are highly recommended for anyone who wants a thorough and professional history of Deere & Company. A second author of special note is Don Macmillan, particularly the 1960–1990 Volume 2 of *John Deere Tractors and Equipment*. Authors of the other material consulted include: Rod Beemer, Randy Leffingwell, David Magee, Barbara Marsh, Faith Hamilton Meyer, Merle L. Miller, James O'Toole, Chester Peterson, Robert Pripps, Brian Rukes, Ralph Sanders, Norm Swinford, and the staff of *Green Magazine*.

Very knowledgeable individuals have assisted this process in various ways. Of particular note is Anna Wolfe, who reviewed the manuscript and improved certain points. Tony Gerber, owner of T&D Restorations in Millbank, Ontario, contributed encouragement and photography. His specialty is restoration of John Deere tractors from this period. Tony has been an encourager and supporting photographer throughout this project. Several retired engineers also reviewed the manuscript, making numerous detailed corrections and providing invaluable personal recollections.

Special thanks to my faithful and understanding spouse, who permitted me the time to "indulge" in this project. For better or worse, I couldn't have done it without you, Angie.

ABOUT THE PHOTOGRAPHY

For the most part, the equipment photos in this book are of tractors and machines that are anything but pristinely restored. Many have oil and grease accumulated over several thousand hours of real field work and still serve their owners. Some are abandoned to nature, parked permanently in pastures and woodlots. Some were never green and yellow, but represent significant points in the history of the agricultural machinery industry. All modern photographs were taken by the author, unless otherwise noted. The vintage photography includes a selection of previously unpublished personal photos from the Hewitt family collection, kindly provided by Anna Hewitt Wolfe. Other vintage photos are from one of the oldest surviving John Deere dealerships, Fred Haar Co., which was established in 1882 in Freeman, South Dakota.

ABOUT THE AUTHOR

John Dietz is a Canadian American freelance writer and photographer living in rural Manitoba. He studied Journalism at the University of Nebraska, Lincoln, in the late 1960s. He worked as a city reporter, rural weekly newspaper editor, and provincial communications officer prior to starting his freelance career in 1980. He is a regular contributor to *Successful Farming Magazine*, a former field writer for John Deere's *The Furrow*, and author of *John Deere Two-Cylinder Tractor Buyer's Guide* and *Classic John Deere Two-Cylinder Tractors: History, Models, Variations & Specifications 1918–1960*.